P27

RICE: THE PRIMARY COMMODITY

The international trade in rice has not been widely studied and is sometimes the subject of sensationalised and misinformed conjecture. This book aims to de-mystify the trade, and reveal its workings and the problems which confront it.

The book outlines the history and cultivation of rice, and the research programmes which have done so much to revolutionise its production in recent years. But countries who buy and sell in the world market change year by year in an unpredictable way due to climatic and political events. This creates a need for brokers who use their expertise to establish contracts between buyers and sellers quickly and efficiently. Trading companies and government agencies are also active. Before the Second World War Burma, Thailand and Vietnam were the major exporters, but since then Burma and Vietnam have experienced difficulties. Fortunately the United States has emerged to challenge Thailand as a world exporter. But the situation in China, sometimes a major exporter and sometimes a major importer remains a cause of concern.

A.J.H. Latham is Senior Lecturer in International Economic History at University College, Swansea. His publications include *The International Economy and the Undeveloped World 1865–1914* (1978), and *The Depression and the Developing World 1914–1939* (1981).

ROUTLEDGE STUDIES IN THE MODERN WORLD ECONOMY

1 INTEREST RATES AND BUDGET DEFICITS
A study of the advanced economies
Kanhaya L. Gupta and Bakhtiar Moazzami

2 WORLD TRADE AFTER THE URUGUAY ROUND
Prospects and policy options for the twenty-first century
edited by Harald Sander and András Inotai

3 THE FLOW ANALYSIS OF LABOUR MARKETS
edited by Ronald Schettkat

4 INFLATION AND UNEMPLOYMENT
Contributions to a new macroeconomic approach
edited by Alvaro Cencini and Mauro Baranzini

5 MACROECONOMIC DIMENSIONS OF PUBLIC FINANCE
Essays in honour of Vito Tanzi
edited by Mario I. Blejer and Teresa M. Ter-Minassian

6 FISCAL POLICY AND ECONOMIC REFORMS
Essays in honour of Vito Tanzi
edited by Mario I. Blejer and Teresa M. Ter-Minassian

7 COMPETITION POLICY IN THE GLOBAL ECONOMY
Modalities for cooperation
edited by Leonard Waverman, William S. Comanor and Akira Goto

8 WORKING IN THE MACROECONOMY
A Study of the US labor market
Martin F. J. Prachowny

9 HOW DOES PRIVATIZATION WORK?
edited by Anthony Bennett

10 THE ECONOMICS AND POLITICS OF INTERNATIONAL TRADE
Freedom and trade, volume II
edited by Gary Cook

11 THE LEGAL AND MORAL ASPECTS OF INTERNATIONAL TRADE
Freedom and trade, volume III
edited by Asif Qureshi, Hillel Steiner and Geraint Parry

12 CAPITAL MARKETS AND CORPORATE GOVERNANCE IN JAPAN, GERMANY AND THE UNITED STATES
Organizational response to market inefficiencies
Helmut M. Dietl

13 COMPETITION AND TRADE POLICIES
Coherence or conflict
Edited by Einar Hope

14 RICE: THE PRIMARY COMMODITY
A. J. H. Latham

RICE: THE PRIMARY COMMODITY

A.J.H. Latham

London and New York

First published 1998
by Routledge
11 New Fetter Lane, London EC4P 4EE

Simultaneously published in the USA and Canada
by Routledge
29 West 35th Street, New York, NY 10001

© 1998 A.J.H. Latham

Typeset in Garamond by Routledge
Printed and bound in Great Britain by TJ International Ltd,
Padstow, Cornwall

All rights reserved. No part of this book may be reprinted or
reproduced or utilised in any form or by any electronic, mechanical,
or other means, now known or hereafter invented, including
photocopying and recording, or in any information storage or
retrieval system, without permission in writing from the publishers.

British Library Cataloguing in Publication Data
A catalogue record for this book is available from the British Library

Library of Congress Cataloging in Publication Data
Latham, A.J.H.
Rice : the primary commodity / A.J.H. Latham.
Includes bibliographical references and index.
1. Rice trade. 2. Rice. I. Title.
HD9066.A2L38 1998
338.1'7318–dc21   97–41056

ISBN 0–415–15153–8

TO DAWN AND TOBY

CONTENTS

List of tables ix
Preface x

Introduction 1

1 Rice and rices 2

2 Irrigation and cultivation 10

3 Mills and milling 20

4 Trade and commerce 27

5 Brokers and traders 35

6 Countries and policies 46
   India 46
   Pakistan 50
   Bangladesh 51
   Myanmar 52
   Malaysia 54
   Indonesia 55
   Thailand 60
   Vietnam 62
   The Philippines 67
   China 68
   North Korea 78

South Korea 79
Taiwan 79
Japan 80
Australia 82
United States of America 83
Brazil 89
Uruguay 90
Argentina 91
Peru 91

7 Trade in the 1990s — 93

Appendix — 99
References — 104
Index — 109

TABLES

4.1	Rice exports (selected countries) 1934–90, metric tons (000)	31
4.2	Rice imports (selected countries) 1934–90, metric tons (000)	33
7.1	World rice exports 1990–6, metric tons (000) (milled basis)	94
7.2	World rice exports 1990–6 (%)	95
7.3	World rice imports 1990–6, metric tons (000) (milled basis)	96
7.4	World rice imports 1990–6 (%)	97
7.5	China's net rice exports 1990–6, metric tons (000) (milled basis)	98

PREFACE

I was asked if I would prepare a text on rice as a commodity in the contemporary world, whilst busy on a book on the history of the international rice trade. That work is not yet finished, but I have managed to seize the opportunity to put this study together in the midst of my other duties as Senior Lecturer in International Economic History at University College of Wales, Swansea. I have tried to write it in as straightforward a way as possible, and I hope it will be of interest to specialists and general readers. I should like to thank Douglas Waller and the late Alan Harper of the London Rice Brokers' Association for their help and encouragement, although they are in no way responsible for any errors or views expressed here. I should also like to thank Morgan Perkins of the US Department of Agriculture for permission to use material from *Grain: World Markets and Trade* in my research. Thanks too to Chris Black of the Rice Growers' Co-operative Ltd of New South Wales for assistance in respect to the Australian rice industry.

INTRODUCTION

Rice is a primary commodity because it is a widely traded basic foodstuff. I have called it 'the' primary commodity because it is consumed in Asia and to a lesser extent Africa and South America by people of low income to whom it forms a large part of their diet. Yet although about half the world's population is dependent on rice, it does not figure in the world's great marketplaces. The Chicago wheat market has no parallel in rice, and the one futures market for rice that does exist, only deals in unhusked US rice. Rice is traded by dealers and governments across the world in what appears a bewildering network of contacts. This study tries to give some insight into the nature of rice as a commodity, and how this vital foodstuff is traded.

1

RICE AND RICES

Rice is a luxury crop. It is usual to assume that it is the food of the Asian poor. But the key to understanding the spread of its cultivation is to realise that it was a preferred grain amongst many other grains and foods which were and are regarded as inferior. These included tubers, millet, sorghum, rye, barley, wheat, pulses and even relatively recent introductions from America such as maize and sweet potatoes. Just as in Europe white wheat bread was regarded as superior to brown bread, oats and rye, so in Asia rice was regarded as superior to other grains. As incomes rose, people in Europe increasingly switched to white bread, and in exactly the same way people in Asia switched to rice (Latham 1994a).

There is linguistic and ethnographic evidence to suggest that the earliest staple foods grown in monsoon Asia were tubers such as yams and taro, and grains such as foxtail and broomcorn millets (Barker *et al.* 1985: 2; Bray 1986: 20). These would have been augmented by other vegetables, meats and fruits.

To much of Asia, rice is a relatively recent introduction. The most ancient archaeological finds, in the Yangzi delta in China, are dated back only 7,000 years (5000 BC). The earliest known remains of rice in India are dated to a period as recent as 3,500 years ago (1500 BC). In Japan rice seems to have been cultivated first in the southern island of Kyushu about 2,400 years ago (400 BC), presumably having been introduced from the Yangzi region, and to have gradually spread northwards.

Millets and tubers continued to be dominant in the Malay peninsula and Indonesia until about 1,200 years ago in the ninth century when they were displaced by rice, and irrigated rice apparently only came to Java 800 years ago in the thirteenth century. What is clear however, is that when rice was introduced to these societies, they did not revert to their previous diet. Rice had arrived permanently (Bray

1986: 9–11). This was because it was preferred to the other crops upon which the people had previously depended. Even then however rice was supplemented by the earlier crops, and indeed with later introductions. In the Tokugawa period in Japan only 50 per cent of the diet of peasants was rice, which was nearly as much as the 60 per cent enjoyed by many people in south-east Asia in the 1970s (Kito 1989: 43). Accordingly, approximately half of people's diet was of items other than rice.

The expansion of rice production continued even during the twentieth century. Although rice has been grown in Korea for two thousand years it was not the diet staple, and Koreans, like the Japanese, actually depended upon other grains and foods. As late as the 1920s only a third of cultivated fields were under wet rice, and the economy actually depended upon dry rice, and field crops like millet, barley and soybeans. These were less vulnerable to drought than wet rice. Rice culture there did not reach its fullest development until the Japanese colonial government carried out a programme to increase rice production in the Korean peninsula. So it was only in the 1930s that rice became the staple Korean dish (Lee 1989: 21; 1990: 55). As for India, Grist noted as recently as the 1980s that in some areas *ragi* (*eleusine coracana*) a coarse millet was the main diet, but when people began to eat rice they were reluctant to return to *ragi* (Grist 1986: 487).

There are two cultivated rices, *Oryza sativa* of Asia, and *Oryza glaberrima* of West Africa, but it is *O. sativa* which dominates commercial usage. Over the years it has differentiated into three subspecies linked to the conditions under which it is cultivated. Indicas originated in the Asian tropics and subtropics, sinica-japonicas in the subtropical and temperate zones, and javonicas in the equatorial climate of Indonesia. Sinica-japonicas are usually known as japonicas although they are now thought to have originated in China. Indicas and japonicas are the two most important. These rices are also divided into 'dry' rain-fed upland and lowland rices, and 'wet' irrigated and 'floating' deep-water rices. Some rices for example can grow in water up to 5 metres deep, rapidly growing ever longer to accommodate rising flood waters.

These rices also have different characteristics when cooked and eaten, with indicas having long thin grains, which do not stick together when cooked, the most favoured types from India to Thailand also being fragrant or scented. They contrast with the more rounded grains of the sinica-japonica varieties, which go sticky and coagulate when cooked, characteristics favoured by the Japanese (Swaminathan 1984: 63–4, 68).

The earliest cultivated forms of rice are thought to have been indicas, developing from wild ancestors in northern and eastern India, northern south-east Asia and southern China. These were tropical varieties, but as cultivation spread, a temperate zone rice evolved in China, which was the ancestor of the sinica-japonica varieties (Swaminathan 1984: 63). Both kinds of rice continued to be grown in China, where they are noted in the earliest Chinese dictionary of AD 100. But as the sinica-japonica types are very photosensitive, they do not flourish in the short days of the tropics, and prefer the longer day lengths of more temperate climes. This is why they predominate in north China, Korea and Japan. They also flourish at higher altitudes than the indica varieties (Bray 1986: 12).

Quick ripening 'Champa' varieties, originated in the state of Champa to the south of Annam in Indochina, and were known 2,000 years ago (AD 100). They could be cropped twice a year. By AD 600 many varieties had been developed, both wet and dry, and early and late. However their yield was poor, although they had low water requirements and could tolerate drought well and were even used as dry crops. By AD 1000 they had spread to south China. Then they were brought to the Lower Yangzi region, even though they were indicas, and not sinica-japonicas like the rices grown there at the time. Soon higher yielding varieties were developed, and by 1250 they were grown across the region (Bray 1986: 22–3).

The colonial powers in Asia had long maintained agricultural research stations (Headrick 1988: 210–31). In the 1870s the Japanese too established agricultural research stations, as part of their thrust for modernisation and development. So it was that in the 1920s when they found the rices in their colony of Taiwan not to their taste, they encouraged the local farmers to adopt new high-yielding japonicas. These commanded double the price of local rices, and were mainly exported to Japan, so the local farmers quickly took to them. But the farmers themselves, and other Taiwan consumers, continued to prefer their traditional indica varieties, and they do so to this day (Bray 1986: 23).

Indicas were carried from India to the Middle East, North Africa and even Europe 3,000 years ago (1000 BC), and they were also taken to East Africa and Madagascar. Javonica strains too arrived in Madagascar, direct from Indonesia. The United States then received rice from Madagascar (Dethloff 1988: 8–9) and south and east Asia. South America obtained rice by way of Europe, and so did West

Africa. However, the African rice *O. glaberrima* is thought to have originated in the Niger delta and spread westwards along the swampy areas of the African coast, where it is the dominant type (Swaminathan 1984: 66).

Rice is a member of the grass family, like oats, barley, rye, and wheat, and it too is an annual. It is thought that apart from wild rices, there are in the region of 120,000 varieties of cultivated rice. Starting in the colonial days of the 1930s, the leading rice growing countries began to collect the various rices in their region. By the 1980s China had collected some 40,000 varieties, India 25,000 and even the United States, by then a major rice producer herself, over 7,000.

These were mostly cultivated varieties, and few wild rices featured in these collections (Swaminathan 1984: 63, 66–7). The numerous varieties are accounted for by the cultivation techniques of the peasant farmers. Rice was harvested using a knife held in the palm of the hand, the heads being cut off by pulling the stalk across the blade with a movement of the finger. Because the heads were cut off individually in this way, it was easy for the harvester to see any heads which were of outstanding quality and put them on one side for next season's crop. Similarly the farmer could pick out early ripening seeds, and those which had flourished in the particular conditions of that growing season. Sickles are now widely used, particularly in India, Burma, Vietnam and Japan, because with the new varieties which ripen more evenly there is no need to cut specific heads (Bray 1986: 19–21, 124, 157; Grist 1986: 99–100, 167–8). One of the varieties still widely grown in Myanmar (Burma) was selected by a local farmer in the manner described above (Chang and Li 1991: 71).

In 1960 the Ford and Rockefeller foundations, with the co-operation of the Philippines government, established the International Rice Research Institute at Los Banos, Laguna, in the Philippines. One of the earliest tasks was to carry forward the task of collecting varieties of rices, as a germ plasm bank. This was to ensure the preservation of existing varieties, and provide material from which new varieties could be bred.

By 1970 12,000 varieties had been collected, and in 1971 it became the central collection of the world's rices. Now the collection includes over 74,700 Asian rices, 1,330 African rices, 2,216 wild rices and 2,073 samples are maintained in the field (IRRI 1994: 3, 37). Thousands of breeding lines with specific desirable characteristics are also maintained there. Many of the remaining 30,000–40,000 varieties have recently been collected. To ensure security from earthquakes and disasters, a duplicate set of seeds is maintained at the US

National Seed Storage Laboratory in Fort Collins, Colorado (Swaminathan 1984: 67).

Apart from establishing a collection of wild rices and cultivated varieties, the other major task of the Institute was to use this genetic material to breed new higher yielding varieties which would be disease and pest resistant. When the Institute was established yields of rice in most Asian countries were almost static. Japan was the exception. There yields had increased from 1.3 tons per hectare in AD 900 to 2.5 tons per hectare in the late nineteenth century when modernisation of agriculture took place after the Meiji restoration. Yields further improved in the early twentieth century due to genetically improved varieties, and better husbandry. Now Japanese farmers obtain 6 tons per hectare, and so do farmers in South Korea, North Korea, Australia and California. Yields everywhere in Asia have been increasing since the 1970s but many Asian farmers even now only produce 2 tons per hectare. These improvements have been due to the development of higher yielding indica rices. Traditional varieties of indica were long stemmed, and in consequence tended to fall over due to wind and rain and the weight of the seed in the heads. This 'lodging' caused much grain to be spilled or wasted, so modern semi-dwarf varieties were developed with shorter, stiffer stalks. *Dee-geo-woo-gen*, a mutant originating in China, was the ancestor of many of these semi-dwarf varieties. The first semi-dwarf indica bred outside mainland China was called *Taichung Native 1*, and was released to farmers in Taiwan in the 1950s. Not only did it not lodge, it was also very responsive to fertiliser, and would grow satisfactorily within a wide range of daylight time. So it became a primary task of the Institute to set up breeding programmes to improve upon the successes of these early semi-dwarf indicas, using characteristics bred in from the rices of many nations (Swaminathan 1984: 69) The semi-dwarf IR8 was the first product of these schemes, which was a cross between *Dee-geo-woo-gen* and *Peta*, an Indonesian variety. Under ideal conditions this could yield 10 metric tons per hectare (Barker *et al*. 1985: 3). It was released in the Philippines in 1966 and revolutionised yields, becoming known as the 'miracle rice' of the so called 'Green Revolution'. Other high yielding semi-dwarfs were developed, such as IR5, IR20, IR22 and IR24. These matured in fewer and fewer days, and responded more and more to good farming. As the number of days required for the rice to ripen shortened, farmers found they could grow two or even three crops a year, provided there was sufficient irrigation and fertiliser. (Swaminathan 1984: 69–70; Chang and Li 1991: 50–5, 68–9; David and Otsuka 1994: 3).

So important was fertiliser in obtaining these yields that in 1975 Grist argued in the introduction to the fifth edition of his classic study on rice that there was no such thing as 'miracle rice', or a Green Revolution. High yields could only be obtained under perfect soil and water conditions, with meticulous attention by the farmer, protection from insects, diseases and weeds, and large amounts of artificial fertiliser. He maintained that the costs of artificial fertiliser, pesticides, fungicides and herbicides were too much for the average farmer, so the huge increases in yield possible under ideal experimental conditions could never be achieved out on the farms (Grist 1975: ix). Ten years later he had modified his views a little, and accepted that under perfect conditions it was possible by continuous cropping to produce a yield of 20 tonnes per hectare. The build-up of pests expected under such conditions did not occur, because although the pests did increase, so also did the predators which fed upon them (Grist 1986: xiii–xiv).

Pests and fertiliser became crucial areas of research as the new high yielding varieties became established. Brown planthoppers thrived on the new varieties, and quite apart from doing damage themselves, also transmitted diseases to the plants. When IR8 was first introduced, only one kind of brown planthopper was known to attack rice plants. But the planthopper quickly develops resistance to insecticides, and can even evolve quickly to feed upon varieties initially resistant to them. Soon another kind had evolved, and was damaging the plants. Hence the introduction of IR36, which resisted four leading rice diseases, and four rice insects, including two types of brown planthoppers. It could be grown in many different environments and soil conditions, and provided good quality grain in 110 days, making it possible for three crops a year to be grown under well-irrigated conditions. IR36 was the result of crossing IR8 with that early Taiwanese semi-dwarf, *Taichung Native 1*, and also a wild species from India, *O. nivara*, which had the crucial property of being resistant to brown planthoppers and other insects. But it was not long before a third kind of brown planthopper had emerged, which could feed even on IR36. So yet another variety, IR56 had to be introduced in areas like the Philippines where this was a problem. Clearly it was to be a continuous battle to find new strains of rice to outwit each new generation of brown planthopper (Swaminathan 1984: 68, 70) Meanwhile IR50 and IR58 were introduced to reduce maturity time by five to ten days (Chang and Li 1991: 68–9). A recent rice brought into general introduction is the high yielding IR72 (IRRI 1994: 19). Although it is a high yielder, a new problem became apparent with the introduction of this rice, that of sheath blight (IRRI 1994: 20). This is caused by a fungus, and particularly causes yield losses where rice is intensively cropped using nitrogen fertiliser to obtain good crops. As the nitrogen inputs increase, so do the chances of outbreaks of sheath blight, which means that to some extent the disease can be controlled by limiting nitrogen inputs. But of course, limiting the nitrogen inputs also restricts the potential yield. The problem seems to arise from the fact that modern rices are shorter-stemmed than traditional rices, are planted closer together, and throw up more shoots. This leads to stems and leaves being in much closer proximity to each other in denser stands, which are hot and humid, conditions which promote fungal growth. Research is continuing to find a solution to this problem, and it will be essential to do so if new varieties are to reach their full potential of 15 tons per hectare (IRRI 1994: 22).

As Grist forcibly pointed out, high yields depend upon high inputs of costly artificial fertiliser (Grist 1975: ix). These fertilisers are derivatives of oil, and had been steadily coming down in price during the 1950s and 1960s, and becoming more available to Asian farmers. But the oil crisis of 1973–5 led to a sharp reversal of this trend, and prices suddenly increased sharply. In consequence attention was concentrated on natural fertilisers. Animal manures, and vegetable wastes had long been used traditionally, but supplies were too limited for the quantities now required to achieve the best results with the new high yielding varieties (Barker *et al*. 1985: 4–5; Bray 1986: 48–50). However, flooded rice soils naturally generate their own supplies of nitrogen fixing agents, and it was upon these that research now concentrated. When rice soils are flooded, various nitrogen fixing agents are triggered, including a blue-green algae which has a symbiotic relationship with the tiny bright-green surface-floating water fern *azolla*. As the water fern dies and decomposes, nitrogen is passed into the soil and nurtures the rice plants. The fertiliser propensity of this natural green manure has been known for centuries in China and Vietnam. So the International Rice Research Institute began programmes to discover how to stimulate water fern growth, and to encourage farmers to increase its use. It is the presence of these natural fertilisers which has enabled Asian farmers to produced up to 2 tons of grain per hectare over many centuries without any use of artificial fertilisers at all. Promoting the growth of these natural fertilisers increases yields without the farmers needing expensive artificial fertilisers (Swaminathan 1984: 69).

The use of insecticides to control pests, and herbicides to control weeds has also increased with the dissemination of the high yielding modern varieties, in order to improve the environment in which they are grown.

However, expenditure on these items has been small in comparison to fertiliser, which is fortunate because of the toxicity which they may generate (Barker *et al.* 1985: 5).

Fish are often produced in paddy fields, and provide a very important diet supplement and source of income for the farmers. The presence of fish actually increases the yield, sometimes by as much as 10 per cent, for they eat weeds, worms, insect larvae and forms of algae which are harmful to rice. By eating mosquito larvae the fish help control malaria. But the existence of fish, prawns and crayfish in the paddy fields restricts the extent to which weedkillers and pesticides can be used. However modern insecticides, fungicides and herbicides have been developed which limits their ill-effects on fish (Grist 1986: 309–23).

Looking to the future, IRRI has now bred a new variety which will yield 12.5 metric tons per hectare in trial. Resistance to pests, and diseases have still to be bred into this new high yielding variety before it is ready for general use. (*Financial Times* 12 January 1995: 39).

China has introduced F1 hybrid rices, but because only the first generation of such rices displays the necessary vigour, the farmer cannot use the seed for his next crop, and has to buy new seed each year from the research stations, making this an expensive strategy (Swaminathan 1984: 71). Nonetheless, impressive gains in yield have resulted from these hybrids which were made possible after the discovery of a pollen-sterile wild rice on Hainan Island in 1970. Available commercially from 1979, these hybrids are now grown widely in Sichuan, Guangdong, Hunan and Fujian provinces. In effect China has experienced a second Green Revolution (Chang and Li 1991: 58–9, 73–6).

About 50 per cent of rice production in Asia comes from irrigated crops, although they only occupy about 30 per cent of the area under rice. Irrigation makes for higher productivity, because a non-irrigated crop such as another grain or pulse can then be grown, then a second crop of rice, all in the space of one year. Dry rice, which only receives rain as a source of water, is grown in both lowland and hill areas, but the yields are low, and the crop is very much subject to drought and weather variation. Average yield is low, from 0.5 tons to 1.5 tons per hectare for upland rice, and little more than that for lowland rice. Clearly research needs directing to increase yields in these marginal lands, and this is now belatedly getting under way (Swaminathan 1984: 71; IRRI 1994: 25–9).

2
IRRIGATION AND CULTIVATION

More than 75 per cent of the world's rice is 'wet' rice from irrigated fields, and over 90 per cent of these irrigated paddy fields are in east Asia. For the most part it is from these fields that the increase in rice production has come in recent years, which has maintained rice supplies and stabilised prices. Nearly another 20 per cent of world rice production comes from lowland areas where the rice is only watered by rainfall. The remaining 5 per cent of world production comes partly from rainfed hill areas and partly from swampy flooded areas. Yields in these areas are low, and in the upland rice areas slash-and-burn farming causes serious soil erosion problems (IRRI 1994: 19–30; see also Mikkelsen and De Datta 1991: 103–86). Although slash-and-burn farming often gives good yields in the first year this is because of the humus which has been deposited by the vegetation, and the ashes of the burnt bush which provides nutrients to the new crop. But these are soon used up or washed away, and the fertility of further crops is greatly reduced. So the people move to another area where the process of cutting the bush down and burning it takes place once more (Bray 1986: 28). In the flooded rice areas in Bangladesh and Burma (Myanmar since 1989), there are losses to bandicoot rats and other pests (IRRI 1994: 30–1). Another floating rice area is in Cambodia near the Vietnam border, along the Mekong river. The level of the river rises in late June, and continues to rise up to the end of September to a depth of about 2 metres, after which the level falls quickly (Grist 1986: 62; Huke and Huke 1990: 17–19; see also Huke 1982: Maps).

To provide irrigation, terracing has been built in many parts of Asia, including Java, Sri Lanka, Japan and northern Luzon in the Philippines. The terraces follow the contours of the land, and sometimes are very narrow and tiny. The water comes from mountain streams and is diverted into narrow channels to ensure the whole area is irrigated. But the problem with terracing is that it can lead to erosion and mountain slides (Grist 1986: 55–6). Nevertheless terracing does make it possible to hold water back which would otherwise just run away, and this can be used to supply crops on land which previously could only grow dry crops, or not be cultivated at all. However it would probably be wrong to assume that terracing was built specifically for rice. Terraces in Luzon are usually allowed to dry out when the main rice crop has been harvested, and other crops are planted, and it is likely that many terrace systems were built before rice arrived there (Bray 1986: 10, 20, 32–3).

Rice is by nature a semi-aquatic or swamp crop, and its response to water is different from other upland crops of the tropics. Whatever the variety, rice gives its best yield with approximately 6 mm of water a day.

When water supplies fall below this level yields are quickly reduced, except with those varieties naturally capable of withstanding drought conditions. This sensitivity to water shortage makes rice farmers without regular water supplies very vulnerable, and makes them reluctant to invest in modern varieties or fertilisers. Even a lack of cloud cover in the dry season will cause problems, as then the land will dry out too quickly, depriving the rice of water and nutrients (Barker *et al.* 1985: 102).

It is clear then that irrigation provides the most effective system for growing rice, and indeed makes it possible for rice to be grown continuously on the same land. This is because the water supplied carries nutrients essential to the successful growth of the plant. The yield increases which have taken place in recent years have largely been associated with improved irrigation. Irrigation means not only supplying water, but also draining water when necessary, and the supply and control of water is essential if high yields are to be obtained. Modern varieties of rice are dependent upon good water supplies, and cannot attain maximum yield without it. Water is more important than the type of soil, and given adequate water, rice will grow satisfactorily on many different soils, and in many different climatic conditions. Successful cultivation requires keeping the crop flooded for most of its growth period. The water engineer therefore is a key figure in improving rice production. River water is the best source of irrigation water because it carries particles of clay, organic fertiliser, and dissolved minerals. Yet paradoxically the actual water needs of rice are no greater than that of any other dry crop, and it is the extra nutrients which rice can absorb under swamp conditions which make the difference (Grist 1986: 41–4).

IRRIGATION AND CULTIVATION

Continued cropping of rice on irrigated land actually raises the fertility to a higher level than the land previously enjoyed, and it maintains this high level of fertility in subsequent years (Bray 1986: 28).

There is no set routine for flooding the rice, because it is grown in different places under widely different conditions. Often the fields are flooded and subsequently drained or left to dry out several times, enabling weeding and the application of fertiliser or manure to take place. Small quantities of water frequently supplied seem to give the best results, especially when modern nitrogenous fertilisers are used. It is particularly important that ample water is provided when the crop is forming the ears of grain. But there seems little to choose between intermittent flooding, and keeping the crop in a shallow depth of water which is then allowed to dry out about a month before harvest. Where little fertiliser is available, the most effective system seems to be to keep the crop in water except at the stage when the plants are throwing up new shoots, and in the period just before harvest. Often the land is left dry between harvest and the next planting, because no water is available due to dry weather. This seems to be beneficial, as it releases nutrients bound in the soil. However, continued submergence after harvest seems to do very little harm (Grist 1986: 47–50).

The original rice farmers grew their rice in swamps (Bray 1986: 29). Subsequently rice has traditionally been grown in tropical rainfall areas, or close to rivers which seasonally overflow their banks. Irrigation systems associated with these rivers depend upon the natural flow along them due to the incline of their channels. Dams are also used to control the water. So rice cultivation depends not so much upon heavy rainfall but upon adequate irrigation systems. Having said this, drainage is sometimes more important than irrigation, especially in swampy regions. This is achieved by enclosing areas with dikes which prevent the water flooding freely onto the land, and enable controlled flooding to take place using sluice gates (Grist 1986: 51–9).

In the monsoon area, the crop cycle starts with the rains. When the soil is sufficiently soggy, the field is turned over. In some cases bullocks are loosed into the field who churn it up with their hooves. Simple bullock drawn ploughs and harrows have also been used since ancient times, and still are. Although they only dig a shallow cut, this is ideal for their purpose, as in this way they do not damage the clay foundation underneath the soil which retains the water. Where there is not enough forage to feed bullocks, heavy iron hoes are used by the farmers to turn the soil to the right constituency before planting. These days small mechanical tillers and tractors are often used. Then the seed is scattered by hand direct into the field, or seedlings transplanted from a seed bed. Depending upon the variety, it takes four to eight weeks for a seedling to grow to about the span of a hand in length, when it can be transplanted. The advantage of growing the seedlings to this size in a nursery means the main field is available for other crops during these weeks. Transplanting encourages the root system to spread, and stimulates the throwing up of extra shoots which will eventually result in more ears of grain. Transplanting is usually done by hand, using the spread of the hand as a spacer, a back-breaking task. But it does mean that regular spacing is ensured, in a way that scattering the seed broadcast does not. The young plants are grown on in the flooded field, and if the water can be controlled, the level will be adjusted to the height of the plants. Weeding may take place if the soil is allowed to dry, and fertilisers may be applied. In the month before harvesting the field is allowed to dry out before reaping takes place. Then the grain is threshed, dried and stored according to local practice (Bray 1986: 42–3, 46–8; see also Vergara 1991: 13–15).

The lowest yields are found where fields are neither levelled nor turned over, and the seed scattered by hand, with rain as the only source of water. No fertiliser is used, no weeding is done, and pests are left to thrive. Large areas of Laos and Cambodia used to be cultivated in this way. By contrast high yields are obtained where the fields are levelled, prepared and fertilised and water supplies carefully monitored through an irrigation system. Selected seed is sown in seed beds to produce seedlings which are planted systematically in regular lines. Repeated weeding takes place, together with applications of fertiliser, herbicides and pesticides. These practices are essential to obtaining the best results from the new high-yielding varieties. They combine the best practices from ancient tradition, plus advances which resulted from research in Japan, and Taiwan in the early twentieth century (Bray 1986: 45), and also British and Dutch colonial research stations as exemplified by Grist's masterstudy (Grist 1986; see also van der Eng 1994: 22–3).

Yield however is not necessarily the criterion by which a rice system should be judged. Low yielding systems may simply reflect the circumstances available. The crucial issue to an individual farmer is output per person-year, not output per hectare, and if there is ample land and water a system of minimum care may well give the maximum output per worker. Systems of meticulous care may simply reflect shortages of land and water. Yield per hectare is also low in the highly mechanised US rice industry where the seed is often sown broadcast mechanically.

Yields are improved both by good practice, and by use of fertiliser. Modern high-yielding varieties depend upon inputs of nitrate and phosphorous fertilisers, and may give less than old varieties without them. But the older varieties do not respond as well to the chemical fertilisers and do better with manure and compost, plus the ashes of the stubble of previous crops. Bean cake was particularly sought after as an organic fertiliser, and a single cake when crushed and powdered was sufficient to fertilise a sixth of an acre of rice seedlings, which could then be transplanted into four acres of paddy field (Bray 1986: 48–9). It was so prized that even in the nineteenth century it was brought in quantities by sea from the soya bean growing areas of northern China to southern China, and also to Japan (Kose 1994: 130–2, 136–7). By the 1930s chemical fertilisers were also being imported to the coastal provinces of China in quantity. But the link between irrigation, modern chemical fertilisers and modern varieties is crucial. The expansion of irrigation networks during the Japanese colonial period in the 1920s and 1930s in Taiwan and Korea was part of such an integrated system of advance, resulting in a near doubling of yields (Bray 1986: 49–50). In many ways this expansion characterised the changes which are now associated with the so-called 'Green Revolution'. With improved irrigation facilities harvest fluctuations are less severe, as they are not so dependent upon variations in the rainfall. In many cases a second or even third crop is grown, as the land can be re-flooded. Under these circumstances investment in new seeds and fertilisers is justified, as success is more certain (Barker *et al.* 1985: 96).

In south and south-east Asia irrigation facilities were greatly extended in the colonial period, with the British, Dutch and French investing heavily in irrigation. In India the great grain growing region of the Punjab was developed in the nineteenth century with irrigation facilities, although wheat was the major crop there, and rice less important. There were many other irrigation schemes in other parts however, on the Ganges and in Madras, many serving rice fields. Further massive irrigation schemes took place in the later days of British colonial rule up to 1941 (Headrick 1988: 170–96).

In Malaya too irrigation schemes were introduced by the British (Hill 1977: 113–15, 178–80). The French also invested heavily in irrigation in the Mekong delta (Robequain 1944: 53–4, 67, 92, 110–12, 220–7, 346). In Java the Dutch too were active in providing

IRRIGATION AND CULTIVATION

irrigation schemes for rice both in the nineteenth and early twentieth century (Furnivall 1944: 323–4; Booth 1977: 33–74). There was also great expansion of irrigated rice land in east Asia. Most of the rice growing areas of Japan were irrigated by 1868, and by the outbreak of the Second World War so were two-thirds of the rice lands in their colonies of Taiwan and Korea. Rice yields were running at an average of 2.5 metric tons per hectare, even though modern fertilisers were not yet widely used. A similar situation existed in the major rice growing regions of South China. But there was less progress in most of the rice growing flood plains of south and south-east Asia, where even by the end of the Second World War irrigation provision had barely started. Yields were 1.5 metric tons per hectare, a ton per hectare below yields in Japan.

Irrigation provision in east Asia continued to grow after the war, and so it did in south-east and south Asia facilitating multiple cropping. After 1960 irrigated areas in east, south-east and south Asia grew at more than 2 per cent per annum, so that the irrigated area increased by over a half in these years. Two major post-war irrigation schemes were the Muda River Project in Malaysia which was completed in 1966–70, irrigating nearly 100,000 acres, and the Upper Pampanga River Project in the Philippines completed in 1975, irrigating 83,000 hectares. By 1988 a third of the total rice areas in south and south-east Asia was irrigated, and accounted for 58 per cent of total rice production in the area and the spread of irrigation has continued since. The major river deltas have the lowest percentage of irrigated area, with the exception of the Mekong delta in Vietnam, Laos and Cambodia, where the French colonial authorities had invested heavily in irrigation structures. There has been little expansion there of irrigation since the colonial period. By contrast growth of irrigated area in Bangladesh was the highest of all regions at over 8 per cent per annum. Total irrigated areas in Bangladesh, Burma and Thailand increased by more than a half (Barker *et al.* 1985: 96–101; Huke and Huke 1990: 18)

In south and south-east Asia expansion of irrigation, and the introduction of modern varieties, fertilisers and pesticides came together (David and Otsuka 1994: 3; Chang and Li 1991: 23). The first areas to gain from the spread of the new varieties were those that were already well irrigated. It was less easy to justify the expenditure on chemical fertiliser that was necessary to obtain the best results from the new varieties in areas only watered by rainfall, because of the risk of drought and crop failure. In other words irrigation was an essential prerequisite to the adoption of modern varieties and methods. There

IRRIGATION AND CULTIVATION

was also a substantial increase in dry season crops from irrigated rice areas in these years which is difficult to document. But in both Malaysia and the Philippines the area under dry season paddy increased substantially, and if the pace of expansion of dry season paddy was slower elsewhere, by the late 1970s dry season rice was grown on a third of south and south-east Asian rice fields accounting for a sixth of total rice production in these regions. Work on improving productivity in unirrigated rainfed areas is only now getting under way. The economic position of farmers in these areas has actually been worsened by the effect on prices of the increased supply of rice from the irrigated areas (Barker *et al.* 1985: 96–7; IRRI 1994: 25–9).

It has now been well established that with improved irrigation and modern varieties, there is a substantial improvement in yield. In these circumstances the modern varieties yield five times more than the old varieties grown without irrigation. Of course provision of irrigation will still raise overall production even if the old varieties are used, simply because more land can be flooded and cultivated, and re-flooding and second crops are possible. But the increased yield obtained when the new varieties are combined with the extension of irrigation means that less land needs to be irrigated to obtain a fixed target of production, lowering capital costs. Alternatively, a fixed capital investment in irrigation will yield a much greater increase in production than can be obtained with the old varieties. Of course, the new varieties also need greater inputs of chemical fertiliser, herbicides and pesticides, so although the yield ratio per hectare is 1:5 between the old varieties and the new, the cost ratio per ton is much smaller (Barker *et al.* 1985: 102–3).

In closing this chapter on irrigation and cultivation it is worth making a few observations on mechanisation in rice production. As has already been discussed irrigation schemes have been associated with rice production from early times, and great ingenuity has been shown in the creation and operation of them. Dams, channels, water-wheels, and various wooden devices have been used. The big problem with these ancient methods is that dams rarely lasted more than one season, and could easily be washed away soon after they were built if the rains were heavy. The wooden devices were not robust, and they too only lasted for one season. The two chief methods of raising waterlevels to feed irrigation schemes were dams, and waterwheels or such devices. The Persian waterwheel was a common device in India for raising water from a river to feed irrigation channels, and it was usually undershot, and powered by two-yoked bullocks. The shadoof

IRRIGATION AND CULTIVATION

was another simple and useful device which could raise water 2 or 3 metres, and was essentially a see-saw device with a weight on one end and a bucket on the other, which one could easily operate by pulling on a rope. In China there were foot-pedal machines which operated paddles, and were small enough to be carried around. Two-person models were common, and even six-person models! More sophisticated ones had a simple gearing mechanism that enabled them to be powered by a bullock. These methods are all still in use, but small modern petrol and diesel driven pumps are also available, and IRRI has designed some simple durable modern pumps (Grist 1986: 51–4).

In the United States powerful pumps are essential for irrigating the rice growing areas. In Louisiana and Texas, private companies sell water to the growers, and use huge power pumps to suck the water from the slow moving rivers flowing south, and pour it into the irrigation channels. In Arkansas a similar system operates, although here the water is pumped from wells. This is true of California, except that the water comes from the Sacramento and Feather rivers, and some of it is simply channelled without needing pumps. In the San Joaquin Valley wells with pumps are the main source of supply (Grist 1986: 65–6).

Mechanisation of cultivation has advanced furthest in the United States, where the rice is grown under water. Australia and Surinam, previously Dutch Guiana, also use advanced methods. Indeed the Murrumbidgee Irrigation Area in New South Wales produces the highest yields in the world, due to good water control, ideal soil, plenty of sunshine, and well managed crop rotation (Grist 1986: 492). Investment in mechanisation is only justified where control of water supply can be guaranteed. The land is ploughed by tractor in the autumn in Louisiana and Texas, and it is harrowed in the spring prior to seeding, but in Arkansas and California ploughing and harrowing all take place in the spring. A loosely broken soil is preferred, so that seed will not be buried as lumps of earth break up. Ploughing and harrowing take place according to local soil conditions, light disc harrows proving useful in wet land. On established fields, the land is flooded to about 18 inches, or 50 cm, which helps lumps of soil to break up. Tractors can operate in the flooded fields, pulling harrows, provided the depth of water is not too great. Then the seed is sown broadcast by mechanical seeders, which throws out the paddy as far as 15 metres. On new fields the seed is usually drilled before flooding. In very large fields seed is sown by aircraft, which sow over 160 hectares per day. The seed is soaked before distribution to help germination, and fertiliser is mixed with it so that both seed and fertiliser are spread together. Airborne techniques were pioneered in the 1920s and 1930s. As the seedlings grow, the land is kept under about 5 cm of water, which is gradually increased to about 12 cm. As harvesting time draws near, the fields are drained. This allows the use of harvesters (Grist 1986: 223–34, 492).

Mechanical harvesting equipment can only be used where there are large fields where the rice stalks are standing upright so that they can be cut easily. Animal powered mowers and reapers of traditional kind were found to be unsatisfactory for rice, as a swathe had to be cut round the edge of the field so that the stalks were not trampled down by the horses or bullocks. In any case the old-fashioned machines could not cope with crops which had fallen over or lodged. Binders proved useful in large fields where the rice had not lodged but specially designed combine harvesters have long since replaced them in the United States. Even crops which have been flattened may be harvested with combines if they go slowly with the pick-up adjusted (Grist 1986: 239–42; IRRI 1994: 5)

Combines of course thresh as they cut, but the paddy needs to be dried before storing. As regards threshing, IRRI originally designed a small portable engine-powered thresher for use in Asia. It was operated by three people, one placing the cut stalks onto the tray, another feeding in into the machine, and a third bagging up the grain (Grist 1986: 242–3). Subsequent developments have concentrated on one person stripper-gatherers with built in threshers (IRRI 1994: 24). The newly threshed grain needs to be dried quickly, or it may absorb damp or dew from the air, and crack. With mechanical dryers, it may be necessary to do this in two or three phases. On large modern farms where mechanised harvesting takes place in less than a month, there may not be sufficient dryer capacity to cope with all the crop at once, but it can be held for up to a week or so before drying without difficulty (Grist 1986: 243–4).

For small-hold Asian farmers mechanisation is a mixed blessing. Most of the operations of paddy production and indeed threshing and drying can be done by hand, and where there is sufficient family labour, it may be the best approach. Machines may be labour saving, but they do not increase yield per hectare, and they bring problems of capital, servicing, spares and fuel. Many are not able to afford the costs involved, particularly where the farms are small. Furthermore machines are designed for particular conditions, and often in Asia, paddy growing and harvesting do not take place under the same conditions year by year. This is particularly so when irrigation and water control systems are bad or non-existent, and water supplies are dependent upon rainfall and river level. Nonetheless these days small tractors, and rotovators are increasingly being used (Grist 1986: 245–52; Huke and Huke 1990: 23).

3
MILLS AND MILLING

Rice traded internationally is rice which is surplus to the requirements of the exporting countries. There are a few countries that regularly export rice like Thailand and the United States, but other countries only enter the international market on the occasions they have more than they need. These countries with temporary surpluses do not find it easy to sell their rice. It is of unknown quality to would-be purchasers, who are therefore reluctant to buy. Their rice has no reputation. This does not mean that only high grade rice enters international trade, for much low quality rice is traded. What is crucial is that the rice conforms to standard grades which the market normally deals in. This can only be achieved by milling, and many temporary exporters simply do not have the milling capacity to produce rice of standard grades. In consequence their rice has to be sold at prices which have been heavily marked down (Siamwalla and Haykin 1983: 34–5).

Market prices are ultimately determined by the attitude of consumers, and plant breeders concentrate on producing varieties with a combination of good quality and yield. The flavour, texture, smell, appearance and ease of cooking are all crucial, and without these good characteristics a new variety will prove unacceptable regardless of its other virtues. So quality is determined by milling characteristics, cooking, eating and processing characteristics, nutritive characteristics, and standards of cleanness and purity (Webb 1991: 89–90).

Rice is first subdivided into the categories long-grain, medium-grain, and short-grain, and there are also sub-groups of glutinous and non-glutinous rice, although these are of minor significance. Long- and medium-grain dominate international trade, with short-grain being significant only for Japan, Korea, Taiwan, and China north of the Yangtze. Glutinous rice is mainly consumed in northern Burma, northern Thailand, Laos and Vietnam.

Elsewhere in south-east Asia it is used for special desserts, and for making rice wine. Very little enters international trade, and the main body of rice traded internationally is non-glutinous (Siamwalla and Haykin 1983: 33; Barker *et al*. 1985: 190). In south Asia, parboiled rice is traded, the rice being partially cooked and dried prior to milling, a process which results in less broken grains. The quality of this South Asian parboiled rice is generally poor, due to the low quality of rice, and elementary equipment. Parboiled rice produced in the United States, such as 'Uncle Ben's' sold commercially in supermarkets, has undergone a similar process but is of high quality. Both the United States and Thailand produce good parboiled rice for Africa and the Middle East. Basmati rices, grown in the Punjab area of Pakistan, are also shipped to the Middle East. They are of the highest quality, being long-grained and 'fragrant' or scented (Barker *et al*. 1985: 190).

Commercially traded rice is graded according to the quality of rice after milling, a crucial feature being the percentage of broken grains. In December 1995 Pakistan basmati was quoted at US$410 PMT FOB (per metric ton full on board) whilst Pakistan 25 per cent broken grain was quoted at US$275 PMT (*London Rice Brokers' Association Circular* 1995 1 December: 3). The nutritional value of the two grades is identical, but the price difference is due to the appearance of the rice, and as it is being used for human consumption this cosmetic difference is vital. Only a small proportion of total world rice production enters international trade each year, as most of it is consumed in the producing countries. The different grades and qualities mean that the small percentage of world production which is traded internationally is separated into even smaller subdivisions for commercial purposes. There is however some substitution of one grade for another according to price (Siamwalla and Haykin 1983: 33).

In the United States which operates its own system of standards one of the criterion for grading is the degree of milling, which refers to the amount of bran removed from the grain. For each grade there are defined limits for the percentage of bran, the number of weed seeds, and damaged, discoloured or chalky kernels, and the higher the degree of milling the fewer of these objectionable items will remain, and the better the price it will bring. Well milled rices are usually preferred by rice consumers, although consumers who are nutritionally aware prefer brown rice because it contains more vitamins and trace elements as the bran has not been removed. It is however chewier to eat, and takes longer to cook, and paradoxically is only favoured in the more affluent markets. The degree of milling is judged by observation and comparison with agreed standards, although there is general agreement that this is not entirely satisfactory and a more scientific system would be preferable if one could be devised (Webb 1991: 113–15). Well milled rice may be milled further in a process known as deep milling or overmilling, and this improves cooking quality up to a certain point when further milling starts to spoil cooking quality. Milling to remove between 4 and 9 per cent of the weight of the grain seems to be the optimal level beyond which deterioration takes place (Wadsworth 1991: 379–82). The highest grade of United States rice, US No. 1 must only contain one weed seed in 500 grams. If there are two weed seeds the rice is of US No. 2 quality, even if acceptable in every other way (Wadsworth 1991: 349; Webb 1991: 108, 113–15).

Rice in the husk is milled before it is traded or consumed. This dehusking was originally done by hand using a wooden mortar and pestle, and such methods are still used locally in both Asia and Africa. More sophisticated versions involve a treadle worked by foot. Rows of mortars are sometimes used, and they can be harnessed to water wheels. Another method involves using a tree trunk cut in two and converted into a kind of grinding mechanism, one part turning against the other. These systems are used essentially for dehusking, and the grain is afterwards winnowed to remove the chaff and bran. Such grain often retains much bran, giving the grain a dark and unattractive appearance. But rice traded internationally must be polished and white, and is processed by modern milling machinery. There are many kinds of small milling plants which are suitable for local production, which produce a milling quality equivalent to that of the large mills. These local mills are well-situated being close to the crop, and because the quality is now so good there has been some decentralisation of milling facilities (Grist 1986: 420–5).

The milling quality of paddy (growing or harvested rice, but not yet milled – from Malay pādīh) relates to the size and shape of the grain, which in turn depends upon the variety of rice, how it has been grown, how ripe it is, and how much sun it has had. If it is overripe it tends to break during milling. Milling quality of the grain is also affected by its age, its moisture content, and the way it has been stored. Paddy may be stored for long periods, but milled rice should not be kept for long periods. Parboiled rice tends to become discoloured in storage unless well dried. Dampness can lead to moulds, and spontaneous combustion, weevils may infest the grain and there may be damage by rodents and birds (Grist 1986: 397–411, 425).

The capacity of small mills may be 10 to 75 tons a day, but large mills can produce up to 1,000 tons. These large mills are usually located at the ports of major exporting countries. They produce rice to conform with accepted milling standards and percentages of broken grain, and are capable of producing rice which has been highly milled and polished. The rice passes through five processes, cleaning, hulling, pearling, polishing and grading (Grist 1986: 427–8; Wadsworth 1991: 347–9; Webb 1991: 100–1). When the paddy arrives at the mill it contains pieces of soil, stones, straw, and all sorts of odds and ends, and these must be removed. This requires the use of a screening or riddling process, and a fan to blow away dust. Then the rice is passed over a magnet to remove any nails or bits of iron. The cleaned rice may then be graded into size of grain by shaking it through a series of ever wider meshed sieves. In Asia paddy is usually ready for milling when it arrives at the mill, and it can then be spread out in the sun for a few hours to ensure any remaining moisture has been driven off. But in America where combine harvesters have been used it may be necessary to dry it first (Grist 1986: 427–9, 444–8; Wadsworth 1991: 349–50). Drying on the farm is usually done by placing it in a bin and passing unheated or slightly heated air over it, but at the mill it may be dried with an air flow from heated air dryers (Wadsworth 1991: 365).

There are four basic systems for milling or hulling. The Engleberg huller uses a screw shaped roller turning inside a horizontal tube. The paddy is fed in at one end and husks, bran and milled rice emerges at the other, the husks having been shelled by rubbing against each other. The disc sheller consists of two discs, the upper being fixed and the lower rotating, both having a coarse emery sandpaper-like surface. The paddy is fed through a hopper into a hole in the centre of the upper disc, and falls down between the two discs, where the movement of the lower disc mills it, and works the grain outwards until it falls through a steel wire mesh guard and is collected. Then it is winnowed with fans and the rice, bran and brokens are separated with sieves, and bagged. Roughly shelled rice from this process which still has most of its bran covering is known as *loonzain* in Burma, or cargo rice. But the rubber band husker produces the highest quality of milled rice. Paddy is carried along a conveyor belt to a grooved iron roller which shells the paddy as it passes below. The husks and grain then fall off the end of the belt into a hopper. The rubber roller husker works by squeezing the grain between two rubber rollers which are moving in opposite directions at different speeds. The process is very gentle and provides high grade milled rice with very few brokens.

Hulled rice still has a covering of bran which must be removed by pearling, using a pearling cone. This involves an inverted iron cone surfaced with emery turning rapidly inside a casing lined with steel wire cloth. Rice is fed in at the top of the casing and is scoured as it works its way down between the casing and the cone. The bran passes through the mesh of the wire cloth and is bagged, leaving the pearled rice to fall out at the bottom. The whitest rice is obtained by passing the rice through a series of these cones, each with a narrower gap between the cone and the wire casing. Broken grains emerge with whole grains but they can be separated and re-mixed according to the percentage of brokens required (Grist 1986: 429–36; Wadsworth 1991: 351–9, 362–3).

For the very highest quality of finish, pearled rice is processed again through polishers. These operate on the same principle as the pearling cone, except that they do not use emery on the cones to scour the grain, but sheep skin or leather hide to buff it. Again several polishers may be used in sequence for the finest results, and even colouring matter can be added at the final polish to whiten it still further. Another system consists of a conveyor belt with a coarse surface which carries the rice to a series of traps or trumbles which polish the grains by causing them to rub together.

Full grains and broken grains emerge from the polishers, and for grading, brokens must be taken out leaving only a specified percentage. A series of sieves or perforated cylinders are used, the holes in ever increasing sizes to eliminate the smaller brokens until the required percentage of brokens is obtained (Grist 1986: 437–8). Discoloured or blackened grains may still be present and these can be eliminated using a photo-electric sorter. The rice is poured in a thin stream before photo-electric sensors which see each individual grain and identify dark ones, triggering a jet of air to blow the bad grain away.

For the American and European markets where a highly polished and almost translucent finish is required, talcum powder in a sugar solution is added at the start of the polishing process. Wheat flour, corn flour or salt, may also be employed in a similar way to enhance appearance. For other markets such as India, turmeric is often used to give a yellow colour, and so is saffron. Oils can be added to help prevent deterioration in storage (Grist 1986: 436–40; Wadsworth 1991: 359–63).

So in a typical mill paddy is poured into a hopper, and passes to the cleaning riddle. Then it goes under a magnet to remove pieces of metal, and is lifted by elevator to the huller, from which shelled rice, husk and bran emerges. It passes up another elevator to a sieve over a fan which winnows it, bran and brokens fall through for bagging, and the rice is carried on up another elevator to the pearling cones. The bran from this operation is discharged for bagging and the pearled rice goes up yet another elevator to the polishers. The polished rice is then elevated again to the classifier, winnowed again, and bagged (Grist 1986: 440–1).

Parboiling of rice prior to milling is very common in India, where it is an ancient practice, and more than half the rice consumed there undergoes this process. Elsewhere in Asia parboiled rice is disliked, although Indian migrants have taken their preference for it with them to all corners of the world. Indian communities in Sri Lanka, Malaya, Mauritius, Madagascar, Guyana and the West Indies continue to consume parboiled rice. The paddy is steeped in cold water, and then in hot water, or it may be gently steamed. Then it is dried before milling. The advantage of the process is that the paddy is hardened, and can be milled with less breakage, a big advantage with large grained rices and softer medium length rices. The process can be done in many different ways, on a small scale at village level and at large plants. The basic process involves the rice being steeped in hot, but not boiling water, in brick tanks for three days. Or it may be boiled for 20 minutes, then transferred to steel containers where it is steamed for another 15 to 20 minutes, by which time the grain is completely turned to gelatine. In more modern systems the rice is soaked and boiled in the same container, and by using higher steeping temperatures, the time involved is cut from three days to about half a day. The rice then is spread out on a drying floor, and the sun evaporates the moisture. A heated floor may also be used for this purpose. Sun drying is preferred, and artificial drying is said to discolour the rice and give it a sour smell, although this disappears on cooking. In India the rice mills use a system known as double boiling where the rice is first steamed for about three minutes then transferred to a concrete pit where it is soaked in cold water for about a day and a half. Then small quantities are taken out and steamed again to cook the rice. This last stage is usually done at night, and the rice is then taken to the drying floor to dry in the morning sun (Grist 1986: 441–4).

Various developments of the parboiling process have improved the nutritional value of the rice, the H.R. Conversion process, the Malek Process and various others. In the first of these, operated in the United States, the rice is cleaned, then tumbled in a trough of churning water to float off any other extraneous matter. The remaining wet paddy is then turned out for about ten minutes to drain it, then steeped in pressurised hot water for about two to two and a half hours. The water is drawn off and the paddy put into a cylindrical rotating vessel, steam introduced and the paddy heated for a short time. Then the steam is cut off and a vacuum applied to dry it. Afterwards the paddy is placed in bins and cool air is applied to dry it. About eight hours later it can be milled in the usual way. In the Malek process the paddy is soaked in warm water for between four and six hours then pressure cooked for about fifteen minutes. The rice is dried and milled as normal, before being packed in bulk or boiled and canned for sale (Grist 1986: 480–1; Luh and Mickus 1991: 70–82).

Parboiled rice may be kept for very long periods, as it can no longer germinate. The oils in the grain are removed by the process, and the grain is sterilised. It even has better resistance to mildew and insects. As previously noted, parboiled rice yields a higher proportion of unbroken grains when milled, and gives a good yield even from poor quality paddy. More nutrients are also retained in the grain (Grist 1986: 445; Wadsworth 1991: 365; Luh and Mickus 1991: 51–3, 65, 84).

For a discussion on handling and storage in the US today see *Rice Journal* 1997 15 June, pages 11–18.

4

TRADE AND COMMERCE

> 'Structurally, the market is essentially a thin residual market'.
>
> (Siamwalla and Haykin 1983: 9)

The rice that enters the world rice market is residual or 'left over' rice, surplus to the needs of the exporting countries. They take what they need for their own consumption, and export the remainder. It is 'thin' because the amounts of rice traded are very small in proportion to the amount of world rice production, as most producers and exporters are also major consumers of rice themselves. Less than 5 per cent of annual world production enters the world market. This would be no problem if the sellers and buyers in the market were the same year by year. But the market is in fact extremely volatile, with sellers and buyers changing all the time, according to the state of their own crops. A bad harvest may suddenly take an exporter out of the market, or even force them to import rice. Similarly a good harvest may make it unnecessary for a country to import rice, and possibly leave them a surplus which they can export. So year by year the participants change, with different buyers and sellers entering the market (Siamwalla and Haykin 1983: 9–13). Sellers and buyers must link together quickly, and as the situation each year changes, finding a suitable partner is a bewildering and confusing process. The search takes time and money so the cost of actually making the transaction is high. For this reason specialist rice brokering houses exist in the major market centres, who make their living from the commission they charge in setting up these deals. There are brokerage houses in the United States, Europe, Singapore and Hong Kong (Siamwalla and Haykin 1983: 34–5). In Europe, brokerage houses exist in Britain, France and Belgium (*London Rice Brokers' Association Circular* 1

August 1955: 4). This is a situation quite unlike that in the world wheat market, where the Chicago market acts as a central market or wheat supermarket. An attempt to set up a futures exchange for rice, the New Orleans Commodity Exchange, ended in failure in the early 1980s (Siamwalla and Haykin 1983: 9, 61; Barker *et al.* 1985: 191–2). Subsequently a rice futures market was established in Chicago but it only deals in US rough rice (see Chapter 5).

Trade in rice dates from ancient times. As early as AD 900–1000 there was long distance trade in rice in China along the Yangzi and other major rivers in the south, availability of water transport and hence cheap freight costs being crucial. By 1200 ocean-going junks were carrying rice to Indochina. But exports gave way to imports, and by 1720 rice was being imported to the Yangzi delta not only from the central Yangzi region but also from Taiwan (Bray 1986: 127–8).

The colonial period of the nineteenth century saw a vast expansion of the international rice trade. Up to 1860, Bengal, under British control, was a major rice exporter, particularly to Ceylon (Sri Lanka), which was also British. Then Burma (Myanmar), also under British control, emerged as a major exporter, much of her rice going to Britain and other European countries. Some of the rice going to Britain was re-exported to the Caribbean and America, the Mediterranean and continental Europe. As the population grew in Bengal, more of their annual crop was consumed there and less was left over for export, so Ceylon had to take Burma rice instead. Demand for Burma rice was also growing in Malaya, where there were increasing numbers of Indian labourers employed on the plantations. Rice was brought down to Singapore and Penang from Rangoon, and redistributed on out to the plantations up-country. So Burma emerged as one of the world's great rice exporters, the Irrawaddy delta being naturally suited for rice production. What was happening there was being paralleled in two of the other great delta regions of Asia, the Chao Phraya delta in Siam (Thailand), and the Mekong delta in French Indo-China (Vietnam). Rice from the latter regions was preferred by the Chinese migrants who established themselves all over south-east Asia. Some of the rice from Bangkok and Saigon came south to Singapore, where it met the rice coming down from Rangoon. From there it was redistributed up the Malay peninsula to the Chinese workers in the tin mines and plantations. It was also transhipped from Singapore down to Batavia in the Dutch East Indies (Indonesia) where there were also substantial Chinese communities in the mines and plantations. Bangkok and Saigon rice also went north to Hong Kong, which like Singapore was a major rice-redistribution centre. From there it was shipped on into China, and even to Japan, Hawaii and California. The Philippines too were major rice importers, the rice usually coming direct from Saigon. Singapore and Hong Kong were the two great market redistribution centres in the trade. An examination of the movement of rice prices in the major centres during this period shows that price fluctuations were synchronised across Asia, revealing an integrated market structure. The availability of cheap water transport, by sea, was crucial (Latham and Neal 1983: 260–80; Latham 1986: 645–63; 1988: 91–4).

This pattern of trade continued in the early years of the twentieth century, through to the 1920s. But then came the depression. There was continued expansion of the area under rice in the three great exporting regions, but as early as 1926 prices began to fall. The year 1928 saw excellent crops. Information supplied by the International Rice Research Institute in the Philippines suggests that the yield of rice can vary as much as 10 per cent according to the amount of cloud free sunlight it receives, which may account for the abundance. 1929 was not quite as good a year but it was by no means bad, but 1930 was again exceptionally good, and harvests remained good until 1933. This coincided with a series of good wheat harvests, and prices of both grains fell drastically as world grain markets were glutted. Rangoon No. 2 (rice) fell from over £0.70 per cwt in 1927, to below £0.30 per cwt in 1933, paralleling a similar fall in Chicago No. 2 (wheat) prices (Latham 1981: 176–8; 1986: 654–6, 663).

With the collapse of rice prices the depression came to Asia. Major rice trading companies failed, and so did agricultural banks. Now the various governments tried to isolate themselves from the depression opting out of the world market. They resorted to import controls and import substitution. The freedom of the world rice market was the casualty. As early as 1921 the Japanese government in response to the rice shortage of 1919–21 had tried to stimulate domestic production whilst keeping prices down. In 1933 they imposed duties to keep out cheap foreign rice and grain, and began a policy of government purchasing to raise prices and increase domestic production. They were soon faced with warehousing problems. In 1939 the government took control of the rice markets, and licensed all brokers and dealers. In the American colony of the Philippines rice control was introduced in 1936, and the government corporation bought paddy, milled it and distributed it. In this way the Chinese brokers and millers were excluded, and prices were controlled to both farmers and consumers. A similar policy was followed in Siam where again the government by-passed the Chinese distributors, buying direct from farmers, and selling direct to consumers, using the profits to set up colonies of Thai farmers free from debt to Chinese rice merchants. In China, import duties were raised to prevent imports, and in French Indo-China, export duties were put in place to prevent exports. Previously Indo-China had supplied China and the Philippines. With export controls and import controls operating simultaneously the international rice trade came virtually to a halt. In British Malaya, rice self-sufficiency became the aim, because the collapse of rubber and tin prices in the depression hit her main exports, and deprived her of the necessary foreign exchange to purchase foreign rice. Cheap land was made available to farmers, and irrigation and seed were provided. When war broke out in 1939 rice prices were regulated and minimum prices offered in some states for rice delivered to government mills. Rice imports became a government monopoly in the national interest. The Dutch East Indies followed similar policies, with imports of rice restricted and domestic production encouraged. When war broke out, a commission was established to buy and sell rice to maintain prices and guarantee supplies (Latham 1988: 99–100; Wickizer and Bennett 1941: 165–87).

So Asian governments, both colonial and non-colonial sought to achieve rice self-sufficiency as a response to the shock of the depression. These policies were to be carried over into the post-colonial period. After the Second World War governments everywhere continued control over rice. Initially Burma, Thailand and Indo-China resumed their roles as major exporters, but when Burma became independent the government held prices to such a low level that incentive for farmers to produce rice was reduced and gradually the country faded out of the international market and moved towards economic isolation. Exports plunged in 1966 and remained low for many years. In Indo-China, the chaos in Vietnam, Laos and Cambodia turned this major exporting region into a major importer. Only Thailand remained a leading exporter, and there too the government remained in control. China however, once a deficit country, became an exporter, using the hard currency obtained to purchase wheat, and so obtain the maximum amount of food calories. Pakistan also became an exporter, but the most significant circumstance in the rice trade in the post-war world was the emergence of the United States as a key exporter, helped by subsidised production and aid-assisted sales. Vietnam was a key beneficiary! (Latham 1988: 100; Siamwalla and Haykin 1983:12–16, 39–41; Barker *et al*. 1985: 188–9, 196–7)

Table 4.1 gives some indication of the contribution of the main exporters to the world rice trade since the depression.

Table 4.1 Rice exports (selected countries) 1934–90, metric tons (000)

	Annual average					
	1934–8	1948–52	1960	1970	1980	1990
Burma	3,070	1,231	1,749	640	653	213
Vietnam	1,320	381	659	218	33	1,624
Thailand	1,388	1,293	1,202	1,063	2,796	4,017
China	16	0	1,153	967	1,376	405
Pakistan	393	73	68	482	1,086	743
USA	71	536	884	1,740	3,054	2,473

Source: FAO *Trade Yearbooks*

Another issue in the history of the world rice trade since the Second World War, has been the impact of the High Yielding Varieties (HYVs) of the so-called Green Revolution. It has already been argued that the key characteristic and weakness of the rice market is that it is a thin residual market. Paradoxically the new varieties with their accompanying needs of irrigation and chemical fertiliser have accentuated these problems and made the market even thinner. The reason for this is that historically the three traditional Asian exporters, Burma, Indo-China, and Thailand relied on their environmental advantages as naturally flooded alluvial plains. These were conditions which were not well suited to the new varieties, because water was supplied seasonally by nature and not easily controlled. Under these conditions of spontaneous inundation it was not easy to apply or regulate the application of fertiliser. So for the most part they continued to grow the more traditional varieties which had evolved in their environment. Burma was an exception, and from 1975 began to expand the area under HYVs, in both irrigated and non-irrigated areas enabling some expansion of domestic rice production without substantial increases in irrigation.

However it was the importing countries, such as the Philippines, Malaysia and Sri Lanka, who took advantage of the new technology to make themselves more self-sufficient, and able to depend more on their own domestic production without importing from abroad. So in the 1960s and 1970s the new HYVs lowered the need of the major importers to buy from the old exporting countries. They were able to move towards achieving the goal of self-sufficiency which they had targeted since the 1930s. That this coincided with the exodus of Burma and Vietnam from the international market confirmed the logic of their policy. In consequence the world market, already a thin residual market, became even thinner and more residual (Siamwalla and

Haykin 1983: 11, 16, 20–1; Barker *et al*. 1985: 195). This could be achieved because domestic rice markets were under strong government control almost everywhere in Asia, including even Hong Kong the apostle of the free market! The food supply was considered too important to be left in private hands (Siamwalla and Haykin 1983: 12).

Table 4.2 gives figures for the main importers.

From Table 4.2 it can be seen that India was a major importer before the Second World War, but reduced imports subsequently. Bangladesh however, previously part of India, continued to be a major importer. Singapore and Hong Kong also continued to import, for they were important urban centres and entirely dependent upon outside rice sources. Malaysia however was a rice producer, and made use of the new technology to try to achieve self-sufficiency although clearly still had some way to go even by 1990. Sri Lanka also continued to import rice, although made good progress towards attaining rice self-sufficiency. But the vicissitudes of the rice trade are shown in the figures for Indonesia which imported 956,000 metric tons in 1970, 2,011,000 metric tons in 1980 and only 49,000 metric tons in 1990. As for the Philippines, they managed to cut imports successfully in the 1960s and 1970s, but had to purchase heavily from the international market in 1990. China too saw considerable fluctuations in imports, made all the more curious by the fact they were also an exporter! As already discussed, Vietnam was an exporter who became an importer because of the exigencies of war. Vietnamese rice production has recovered in more recent years, enabling a reduction in imports. Probably the country which was most successful in achieving self-sufficiency was Japan! A general examination of these figures does indicate the inconsistency of the world rice market, with major fluctuations year by year as importers requirements changed.

Another factor affecting the world rice trade since 1945 is the fact that wheat consumption has been growing in many of the major rice consuming countries, and they have been importing increasing quantities of wheat. This is not necessarily surprising as in some parts of the world rice and wheat production overlaps. In northern China wheat is the main crop, whilst in southern China it is rice, but in central China from the Yangzi valley to the Huang He (Yellow) river, both wheat and rice are grown, and consumers switch between each grain. They also eat them conjointly as, for example, rice with noodles, and congee (rice porridge) with crispy breadsticks. The same is true of parts of central India, between the predominantly wheat growing North and the predominantly rice-growing south. Here rice and chapattis are eaten together (FAO 1955: 5 para. 32). Wheat

Table 4.2 Rice imports (selected countries) 1934–90, metric tons (000)

	Annual average					
	1934–8	1948–52	1960	1970	1980	1990
India	2,094	779[1]	698	582	19	66
Bangladesh	–	–	323[2]	510	548	380
Japan	1,757	525	174	18	13	18
Malaysia/Singapore	718	511	641[3]	366	167	330
Singapore	–	–	–	293	188	220
China	704	127	27	5	130	62
Hong Kong	522	163	369	344	359	374
Sri Lanka	530	425	528	544	168	131
Indonesia	280	383	962	956	2,011	49
Philippines	37	92	0	0	0	592
Vietnam[4]	–	–	21	891	451	27

Source: FAO Trade Yearbooks
Note:
[1] Excludes Pakistan and Bangladesh from 1948
[2] Figure is for Pakistan which included Bangladesh
[3] Includes Sabah and Sarawak from 1960
[4] Includes Cambodia (Kampuchea)

prices have dropped relative to rice, because technological advance has made faster progress amongst the wheat producers than amongst the Third World world rice producers. As a result wheat consumption has risen in many rice-producing countries, including Bangladesh, Vietnam, Indonesia, Malaysia, the Philippines, South Korea, China and Japan. (Barker *et al.* 1985: 196–9). One underlying issue is the fact that rice is a preferred grain amongst an array of other grains like millet, barley, rye, wheat, maize, beans and potatoes (Latham 1994: 18–19, 23–4) As noted in Chapter 1, millet is the main diet in parts of India, but when millet eaters become accustomed to rice they are reluctant to go back to eating *ragi* (Grist 1986: 467; see also FAO 1955: 5 para. 29). What seems to be happening is that wheat has been displacing the other inferior grains which make up the basic diet, leaving rice as the preferred luxury grain. It seems clear that rice will therefore continue to be a major food grain, and because of the inability of producers to be entirely self-sufficient despite new varieties and methods, international trade in rice will continue.

5
BROKERS AND TRADERS

Central to the movement of rice as a commodity, are the brokers and traders by whom the business is conducted. Much rice used to be bought and sold in government-to-government transactions. In the difficult years of the 1930s governments intervened to control the rice trade in countries from Japan to the Philippines, Indonesia and Malaysia, often aiming to eliminate the trading role of millers and agents. The coming of war only reinforced the need to control this crucial strategic commodity (Wickizer and Bennett 1941: 165–87). After the war widespread government control continued, with more than half international trade being conducted on a government-to-government basis (Barker *et al.* 1985: 11, 192–3). There has often been a deep seated mistrust in official quarters of private traders, and a reluctance to leave something as basic as the nation's food supplies in their hands. The fear is that they might be able to profit from their control of essential food supplies, and indeed be in a position to hold the people to ransom. Rice riots bring governments down! It has been felt that traders should not profit from people's essential food. Rarely has it been understood that the skills of private brokers and traders will ensure that supplies are secured effectively from volatile markets, and that the brokers and traders depend for their livelihood by so doing. Their skills at arbitraging rice are likely to be more effective in securing supplies at the best price than the services of well-intentioned civil servants. These days government-to-government deals are less important than they were, although still about half of annual trade is done in this way. However, these deals can be complicated and involve much more than a telephone call between civil servants. For example, sales may be held up by lack of authorised means of payment, or they may be linked to barter agreements on fertiliser, or even delivery of military aircraft or warships. In Singapore and Hong Kong, where the trade was largely in government hands and controlled by civil servants, the trade has in recent years mainly reverted to private hands, with regular trade taking place in the high grade rices, particularly Thai 100 per cent fragrant and similar qualities.

In understanding the way in which trade operates it is important to distinguish between brokers and traders. Brokers do not trade in rice themselves, but instead take a commission on deals which they arrange for their clients (see Sewell 1992: 218; Roche 1992: 141). The point has already been made that because much of the rice trade is a trade in surplus amounts of rice which have suddenly come onto the world market, search costs for finding buyers is high. Conversely, countries finding a sudden shortfall find it difficult to quickly locate dealers with supplies to sell. Hence the brokers have a vital role to play, and they make their living by their intimated knowledge of the market. The clients may be countries, or trading companies, and they may also operate for millers and processors who use rice for commercial products such as breakfast foods. As brokers, they enable both purchasers and sellers to preserve their anonymity, which may be important for political or commercial reasons. Trade is a private matter, and one only has to recall the treatment of millers and dealers in Asia in the 1930s to understand why traders prefer to keep the details of their transactions to themselves.

Because of London's historic position as an international market and centre of a far-flung empire with important Asian interests, the London Rice Brokers' Association has long held a central position (Latham 1986: see graph 659). Amongst their other functions they provide a monthly circular of market intelligence (see Appendix).

The Thais, who are major suppliers, are essentially millers, who export rice by delivering it to ships on a FOB (free on board) basis. They are not usually involved in actually shipping rice themselves, although they do occasionally take on the role of shippers by engaging the shipping and selling on a CandFFo (cost and freight free out) basis, with Iraq, Nigeria, Malaysia and others as destinations. But by far the majority of Thai exports, which amount to about 5 million tonnes, are sold through brokers such as Jacksons (UK) or Creed (US). The rice is not sold directly to the international trading companies, such as Glencores (ex Richco), Continental Grain and Cargill (ex Tradax), etc., although the Thais somewhat confusingly refer to them as 'brokers' which they are not. The Thais belong to the Thai Rice Exporters Association (see Appendix for address), and many of them have their headquarters in the traditional centre of the rice trade in Bangkok, the Songwad Road wharf area in Chinatown by the Chao Phraya river. There occasionally you can still see Chinese labourers shouldering blue-striped jute bags between lorry and ship. These B Twill jute bags were traditionally used in the trade, although they are now largely replaced by the cheaper polypropylene 50 kg bags. This manhandling emphasises the low technology of the transport facilities in the trade, with consequent inefficiency and high costs. Nowadays however most of the loading and general handling of rice takes place further down the river at Kohsichang.

These Thai traders have been discussed in sensational terms by journalist Dan Morgan who talks of the Thai rice trade being dominated by 'six tigers' who operate a clandestine monopoly of the rice trade around the fetid Gulf of Siam (Morgan 1979: 293). However, this menacing image is difficult to recognise by anyone who has actually visited the quiet and unpretentious headquarters of the Association at 37 Soi Ngamduplee, and talked to the genial middle-aged gentlemen who work there! The fact there are more like fifty members than six also rather spoils the image! The two leading Thai companies are Soon Hua Seng Co. and Capital Rice Group, and there are in practice a dozen leading exporters and another handful of smaller exporters. Chaiyaporn which used to be the leading company is no longer quite so prominent, although still in the leading half dozen. In stark contrast to Morgan, Sewell stresses the substantial expertise of the Thai traders, their knowledge of markets and qualities, and the fact that deals are done by word-of-mouth (Sewell 1992: 219–20). However, the miller-exporters buy their supplies of paddy from local suppliers on forward contracts, which can cause problems when rice prices are rising. When this happens the local suppliers may sell their paddy for the current high price, and not supply rice at the lower price of their forward contract. In these circumstances the millers may find themselves having to buy at the high current prices to mill rice to meet their export obligations, even though the price they have agreed to sell at may be less than the price they have to pay for the unmilled rice. The best intentioned of merchants can find themselves in difficulty in these circumstances! There were in fact serious defaults on contract in Bangkok in January 1988, on a rising market, which created a serious loss of confidence by foreign buyers. As the *London Rice Brokers' Association Circular* commented 'If exporters expect to default when the market goes up, how can they expect their buyers not to default if the market goes down!' (29 January 1988: 1, 3). Subsequently the position has improved.

The international traders, such as Cargill, Continental Rice and Glencores, are essentially merchants. They buy rice in the country where it originates, charter shipping, and sell the rice in the country where the cargo eventually arrives. Usually the rice is sold while the cargo is at sea, but increasingly the rice goes into a warehouse in the country it has been sent. The Thai exporters used to predominate in the West African trade, but do so no longer, because the trading companies are taking numerous vessels afloat unsold, and they then break the cargoes up at their destination into smaller quantities for local buyers. However the Thais are still very active in the Middle East. Unlike brokers, the traders exist to buy and sell rice at a profit, and in this they build up an extensive knowledge and experience of clients, not dissimilar to the expertise of the brokers. But unlike the brokers they commit themselves to buying and selling, and so take on the trading risk of gain or loss upon themselves. Although they may use brokers to buy and sell on their instruction and pay the appropriate commissions, ultimately they and their bankers shoulder the responsibility for any losses, and the satisfaction of any profit.

Apart from Thailand, the other major suppliers in today's rice trade are the United States of America and also Vietnam, which has recovered from the disturbances of the recent past. US private companies supply Latin America, Africa, the Middle East and Europe.

US exports to Brazil are often as unmilled rough rice, but Peru and the Caribbean Islands take packaged high-quality well-milled 4 per cent broken rice. For the European market, because of EEC import duty on milled white rice, much of the exports are as brown rice or parboiled brown rice, which is then milled in Northern Europe, and sold on to the supermarket chains. Latin America has taken increasing supplies from Vietnam and Thailand, and even though duties have been raised to restrict this flow, the trading companies have found ways of getting round these difficulties. Most of this trade is by the private traders, although some of the trade to the Caribbean and elsewhere is done by tender through US government export support programmes.

For many years the world trade in rice was about 14 million tonnes, with Thailand's best year being about 5 million tonnes, and Vietnam 2–3 million tonnes depending on the crop, with Pakistan providing 1 million tonnes. In 1995 however, the trade increased to 20–1 million tonnes because of crop failures in Indonesia, China, the Philippines and elsewhere.

Private companies are also very active in handling the exports of the United States of America. Companies exporting rice from the US in recent years have included Andre, Louis Dreyfus, Comet Rice, American Rice, Cargill, Continental Grain, Balfour Maclaine and smaller companies such as the Gulf Pacific Rice Co. Creed Rice is a

US broker who issues a weekly market report (see Appendix for address). (See also Roche 1992: 136–7; Broehl 1992: 840.) The man who did much to establish Continental's rice activities is said to have been Raphael Totah, an Egyptian (Morgan 1979: 192 fn, 293, 96, 97). Totah apparently was of Jewish origin, and left Egypt for the US after the revolution in the 1950s which brought Nasser to power (Sewell 1992: 220; Roche 1992: 137). However, this was some time ago and Salvadore Amran has been their rice 'guru' for the last twenty or so years.

In Europe, Britain has been a key figure in the international rice trade from the nineteenth century when importing and trading companies like Steels and Bulloch Bros. were still in operation (Latham 1986: 651). This historical position is still represented by the London Rice Brokers' Association as has been discussed above. However, it must be made clear that the LRBA has nothing whatsoever to do with trading, Jacksons merely providing contracts for trade, and a monthly report. They were the only big operator in London until the new Marc Rich Investments company opened recently. Charles Wimble still operate, basically dealing with intra EEC trade, and one or two other companies handle the odd cargo.

London Rice Brokers' Association contracts are accepted throughout the world. The history of these contracts goes back to the mid-nineteenth century when it was found necessary to have a system of standard contracts, with standardised qualities and a system of arbitration. Arbitration was needed to avoid expensive litigation so that commercial disputes could be settled quickly and fairly within the trade by experts. Arbitration now comes under UK Arbitration Acts 1950 and 1975 and is also covered by international arbitration agreements. Many countries have signed these agreements, and it is possible to go to a court of law in a country which is a signatory with an arbitration award and apply for a court order for enforcement should this become necessary. An arbitration award is therefore much more useful than a court order which may not be recognised, because an arbitration award has automatic recognition (*LRBAC* 31 October 1989: 1).

Rice trading on the continent is conducted from Paris and Geneva, but importing into Europe is controlled mainly by the mills who work brown rice and pack it for distribution. The trading however is more for world destinations beyond Europe, the traders having merely established their domicile on tax and cost grounds.

The French firm Riz et Denrees was established in the 1970s, and became a major trader, but ceased trading due to over exposure in

Brazil. The 1980s saw more European companies become involved, such as Richco and Sucden, but Richco is now Glencore Denrees, and Sucden closed their rice department in the early 1990s after financial troubles arising from other commodities. Other companies are Global, ORCO, CIC and Real Trading (see also Sewell 1992: 220; Roche 1992: 136–8).

Dealing in rice is technically very simple, as is shown by the fact that much dealing in Thailand is by word-of-mouth. However, there is now urgent need for written contracts within Thailand. In particular, futures markets play little part. There was a rice futures market in Rangoon from about 1900 although there were no premises and the dealings were done on the open pavements of Mogul Street (Latham and Neal 1983: 274). Although rice futures are said to have been traded at the Dojima rice exchange in Osaka from the early eighteenth century, it was closed in the 1930s after a series of scandals. There was an attempt to open a rice futures market in New Orleans in 1981, but it achieved little and was transferred to Chicago in 1983, and closed in 1984. Another attempt was made in Chicago in 1986. However this too failed in 1991. In 1992 a further attempt was made, and rough rice futures, and a rough rice option was offered. These are still operating on the Mid America Commodities Exchange (see Appendix). There was also an attempt by the London Futures and Options Exchange (Fox) to set up a futures market in 1990 but this came to nothing (*LRBAC* 30 November 1990: 1).

Of the various reasons for the absence of futures markets, the first and most often quoted one is the vast number of different qualities of rice. To operate a futures market a single, readily definable, identifiable and uniform unit of trade is required. But in rice there are long-grains and round grains, fragrant and non-fragrant grains, glutinous and non-glutinous varieties. There is rice in the husk, milled rice, parboiled rice, untreated rice, and many different mixes of broken and unbroken grain. This is because there are many different sources of rice, and many different consumers of rice, all with their own preferences and prejudices. Quite apart from this lack of a standard item, there is the question of the availability of funding for a futures market, and whether traders and their bankers wish to be involved. The Thai traders who are such important suppliers of rice to the world market, are content with the situation that exists, and it is to be noted that the two recent attempts to establish futures markets have been in the United States and Britain, rather than in Thailand.

To emphasise the difficulties, it is worth recognising that the ventures in Chicago and London involved quite different qualities of rice! The Chicago Rice and Cotton Exchange traded in rough rice, unmilled paddy. The trading unit was 2,000 cwt of US No. 2 or better long-grain rough rice. The delivery points were twelve designated counties in eastern Arkansas. But the London market intended to have a trading unit of 500 tonnes of milled Thai 100B rice current crop in 50 kg polypropylene or jute bags. One futures system aimed at the pre-milling market, and the other at the post-milling market! The London contract also specified that US 2/4 rice bagged similarly could be substituted for Thai rice, if there was a shortage of rice at the port of Kohsichang in Thailand. The rice was to be delivered FOB (free on board) in Kohsichang/Bangkok, or FAS (free along side) in Houston, New Orleans and Lake Charles (Roche 1992: 192–206; *LRBAC* 30 November 1990: 1).

From this discussion emerges the unsophisticated nature of the rice trade. Much of the rice comes from the developing world, and much of it is consumed there. It is not surprising therefore that methods of production, marketing and trading are of a very simple kind involving very few technological inputs. Production and marketing in the United States is an exception to this general picture. But reference has been made to sacks being handled in Bangkok, and the fact that the proposed London market was hoping to deal in consignments of rice in 50 kg jute or polypropylene bags brings this point to our attention. In fact such 50 kg and 100 lb bags are common sizes, although most US mills can provide any specified type of packaging and labelling (Roche 1992: 99). Most of the international trade, including US rice, is of bagged cargo of this kind, and it is transported in smaller, older ships than the bulk carriers employed in the international wheat trade (Sewell 1992: 218; Morgan 1979: 295). Until recently vessels were 8,000–12,000 tonnes, but in the last few years they have been replaced by 15,000–20,000 tonners, because the smaller 'tween deckers are now too old. These ships however are suitable for minor ports with antiquated handling facilities. Once again one is reminded of the Chinese labourers at the Songwad Road wharfs shifting the bags on their backs!

Thus another problem for rice traders is securing suitable shipping arrangements. As many of the ships are small and old and plying less developed routes this is another source of difficulty, inefficiency and expense. As has been stressed before, the developing world inadequacies at all levels of the trade imposes extra costs upon the final price of rice. Buying the rice is one problem, securing the shipping and insurance for it is another! This is brought out by looking at the *London Rice Brokers' Association Circular*. As recently as 1995 a situation of near chaos existed in the shipping of rice in Asia. This was due to the sheer volume of rice being handled due to crop failures in Indonesia, China, the Philippines and other countries. The volume of world trade rose by a half from 14 million tonnes to 20–1 million tonnes. Fortunately India had enormous stocks for export due to good monsoons, going from an annual export of mainly basmati of about 200,000 tonnes to some 3 million tonnes of lower grades, all in the one year. This caused huge logistical problems, but in the circumstances the Indians managed extremely well. In February a long line of vessels was awaiting to berth and load at Bangkok (*LRBAC* 1 February 1995: 3) and in India vessels were taking twenty days just to obtain a berth, before any loading could take place. Meanwhile in Vietnam there were fourteen vessels in port, all loading slowly, including five for Indonesia, one for Mombassa, one for Dar es Salaam, one for Cuba and one for China (*LRBAC* 1 March 1995: 2, 4). By May the situation was dire. Douglas Waller, the long-time Secretary of the Rice Brokers' Association and Editor of the *Circular* (now retired), commented that it appeared as if all the usual rice carrying ships, vessels capable of carrying 8,000–18,000 tons of bagged rice, had been sunk or scrapped! However, as he pointed out, the shortage was really due to the fact that ships were being long delayed both at loading ports and discharging ports. Many ships were delayed in loading at Yangon (Rangoon), and various Indian ports, and over two dozen were held up at Hochiminh (Saigon). To make matters worse, slow discharge in China and Indonesia was also keeping ships out of action. Voyage times in consequence had trebled, forcing freight rates up by 50 per cent (*LRBAC* 1 May 1995: 1). To make matters worse, by the end of May delays were also becoming a problem at the major exporter Bangkok, because of a shortage of lighters necessary to get rice downriver to the ships in port, a problem compounded by public holidays and rains (*LRBAC* 31 May 1995: 3). In June there were still delays at Yangon, and a long line of ships waiting to berth, and the whole situation was being made worse by the rains, and at Bangkok the situation was much the same, with a continuing shortage of barges (*LRBAC* 30 June 1995: 2–3). By August there was a lack of supply from Thailand, Vietnam, Pakistan and Burma, countered only by large amounts being made available from the Food Corporation stock in India. But although the supplies were there in India, transport was not. Railcars were not available to take the rice to the ports, and in any case the infrastructure of the ports in general was not capable of handling the amounts being shipped. So there were big queues of ships in most of the Indian ports waiting to start loading and the arrival of the monsoon was adding to difficulties. At the same time there was extensive flooding in China in Hunan and Jianxi provinces causing damage to crops and warehouse stocks. In Thailand loading continued to be slow because of a clampdown on illegal immigrants, which caused a shortage of stevedores, and hence the lighters which they handle (*LRBAC* 1 August 1995: 1–3). The damage in China appeared to be extensive, and the large supplies from India would have been very useful were it not for the continued problems in the logistics of shipping from all the Indian ports. Shortage of rice was not the cause of the difficulties, but the inability to get the rice to the ports and to load it quickly. There were large lines of vessels in most ports, and it was taking up to two months to load each one, resulting in enormous demurrage bills. In Thailand loading continued slow with the immigrant controls restricting the supply of dockers, and havoc caused by the rains. Parboiled rice was in short supply because it could not be dried outside due to the rain (*LRBAC* 1 September 1995: 1–3). The costs of the delays in India are eloquently displayed by an entry in the October issue to the effect that four or five vessels were on their way or had arrived at Dakar in Senegal bearing Indian broken rice. They had loaded rice in India at around $220 PMT FOB but with demurrage at loading and freight costs added, trader prices in Dakar were more than $300, an increase of $80 or over 36 per cent. The costs to consumers in the developing world of developing world inefficiencies are plain for all to see! (*LRBAC* 2 October 1995: 1). Meanwhile China had become a major importer of rice, presumably due to the extensive flooding earlier reported. But whilst India had the rice to supply, they continued to be unable to move it and by November there were thirty-seven vessels delayed at Kakinada, twenty plus at Kandla and sixteen at Bombay (*LRBAC* 1 November 1995: 2). This situation continued into December and seemed likely to continue for the next two or three months. At Kandla there were loading delays of 40 to 50 days from arrival date, and at Kakinada loading dates could not even be given. To make matters worse, and to add to the railway difficulties, new restrictions on the legal weights for road trucks had been introduced. However, not all the problems in the logistics of rice supply were confined to Asia: in the United States farmers were reluctant to sell paddy as it was the duck hunting season (*LRBAC* 1 December 1995: 2, 4)!

There is no doubt that the *London Rice Brokers' Association Circular* brings one closer to the heart of the rice trade than most other published sources. Quite apart from the news of the general situation, it also gives prices current at the available centres. For example, the September 1995 issue gives rice prices in India packed in 50 kg single jute bags FOB as:

PR 106, 5 per cent brokens	US$305 PMT
PR 106, 10 per cent brokens	US$290 PMT
PR 106, 15 per cent brokens	US$280 PMT
White rice non-basmati 25 per cent brokens	US$255–60 PMT
100 per cent broken rice	US$215/220 PMT

These can be immediately compared with the situation in Thailand where the following prices of rice in 50 kg polypropylene bags are given:

100 per cent B unbroken	US$360 PMT
5 per cent broken	US$350 PMT
10 per cent broken	US$346 PMT
15 per cent broken	US$335 PMT
25 per cent broken	US$320 PMT
35 per cent broken	US$310 PMT
A1 Special	US$293 PMT
A1 Super	US$292 PMT

Parboiled rice was also available at:

PB 100 per cent sortexed	US$370 PMT
PB 5 per cent broken sortexed	US$365 PMT
PB 100 per cent unbroken	US$360 PMT
PB 5 per cent broken	US$350 PMT
PB 10 per cent broken	US$344 PMT
PB 15 per cent broken	US$340 PMT

In Vietnam the following prices were given for government guidance prices FOB stowed Hochiminh:

5 per cent brokens	US$335 PMT
10 per cent brokens	US$325 PMT
15 per cent brokens	US$315 PMT
25 per cent brokens	US$300 PMT
35 per cent brokens	US$290 PMT

In the United States current quotations were:

US long-grain 2/4	US$16.50 per 100 lbs FAS Gulf Port
US long-grain 5/20	US$14.50 per 100 lbs FAS Gulf Port
Long-grain brown rice 2/4/75	US$15.00 per 100 lbs C+F Rotterdam in bulk
Parboiled brown 1/4/88	US$16.00 per 100 lbs C+F Rotterdam in bulk

Month by month similar quotations are given, showing the general movement of prices, which can be used for general guidance. It should also be noted that these quotations also indicate the various qualities generally traded (*LRBAC* 1 September 1995: 2–4).

Other important sources of information on the rice trade (see Appendix), are the weekly *Creed Rice Market Report*. The US *Rice Journal*, is also very useful on general matters relating to the US trade, and their annual *International Rice Industry Guide* is particularly useful as a handbook of important company names and addresses. For more general information there is also the United States Department of Agriculture (USDA), Foreign Agricultural Service, *Grain: World Markets and Trade*, from the US Dept of Commerce.

I am grateful to the late Mr Alan Harper (d.1997) of Jacksons for his helpful comments although he was in no way responsible for any errors or opinions expressed here.

6

COUNTRIES AND POLICIES

The Second World War brought great changes in the International Rice Trade, for there had been devastation and disruption in the three great rice exporters of pre-war days, Burma, French Indo-China, and Thailand. Of these, only Thailand was able to fully return to her position as a major supplier of world needs. Fortunately the US was able to fill the gap caused by the decline of Burma and French Indo-China (Latham 1988: 100–2).

India

In the nineteenth century India exported rice from Bengal, but gradually exports declined and they began to import from Burma, their imports growing substantially in the early years of the twentieth century (Latham and Neal 1983: 262–3). There was a major famine in Bengal in 1944, and the British colonial government introduced a 'grow more food campaign'. They tried to encourage better farming practices using posters and slogans, but without much effect. After independence from Britain in 1947 the Community Development Programme was introduced, which provided an administrative structure down to village level. The Community Development Research Centre was at the core of the system, and the Community Development Organisation became a Ministry. The aim was to improve the general health and education of the people, and make big increases in food production. Food production did rise, cutting rice imports, but in the mid-1950s food production levelled off, and in 1957–8 large imports were needed. So in 1961 the government tried an experiment called the Intensive Agricultural District Programme (IADP), which operated in one district in each of seven different states. Economic and technical innovations were co-ordinated, and improved practices were introduced by village level workers, using better seeds, fertiliser, pesticides and weeding. This led to the Intensive Agricultural Areas programme (IAA) which was introduced in 1964–5. It expanded the area of operation, but did not have the financial backing of the original scheme. Despite these schemes, rice production stagnated.

Good weather gave favourable crops in 1964 but severe droughts in 1965 and 1966 led to huge imports. So the High-Yielding Varieties (HYV) programme was adopted in areas where there was good irrigation. Managers were provided, who were supplied with HYVs and fertiliser. Output rose quickly, and imports began to fall from 1966. By 1970 the food situation was much better, and policy switched to integrated rural development, which aimed at achieving more equal incomes for people in the regions. But 1972 saw more bad weather and rice production fell, and India had to import during the world food crisis of 1973–4. In the mid-1970s there was a shift to a Training and Visits System assisted by the World Bank. Agricultural extension agents toured villages every two weeks training farmers. This was particularly useful in introducing HYVs. The food shortages of these years put the question of income equality into the background. But from 1978 India was able to export rice again, and by 1980 distribution, health, education and family planning emerged as central policy issues. In parallel with the attempts to improve production, India has tried since independence to provide its people with foodgrains at affordable prices. Famine relief by direct distribution gave way in the early 1950s to rationing and fair price shops. The Food Corporation supplied the fair price shops, obtaining much of its supplies from the domestic crop. But it also obtained rice at concessionary terms from the USA PL480 programme, and it bought directly from the world market. To keep prices down, in a strategy adopted elsewhere in Asia, zones were established, and food could not be moved from one zone to another without permission. This held down prices in rice surplus areas, enabling the government to buy more cheaply. The rice could then be moved to food deficit areas at a lower cost than if the free market had operated. The snag was that the surplus areas were the areas which had the resources to provide increased production, and farmers there were discouraged from producing extra rice by the fact the scheme held down prices. However the zones were only imposed in years of shortage. The success of the development schemes meant that between 1978 and 1982 India exported rice. In the 1980s there were further increases in output, particularly in the southern and eastern states. In the 1990s irrigation schemes were begun in Kerala. These years saw government spending on irrigation, drainage, and the extension of HYVs to dry rice areas. Money was also spent on marketing, distribution and cheap credit facilities. Incentives were provided for farmers by increasing the procurement price of paddy each year. Yet paddy prices fell in real terms because of inflation. But as less than half the crop was taken by the government, the remainder could be sold at market prices giving farmers inflation related earnings. The government also subsidised fertiliser supplies. Despite increased production there was still widespread malnutrition. The Public Distribution System (PDS) controls the price of rice and foodgrains, ensuring fair distribution and stable prices. The Food Corporation of India acts for central government and some state governments, and procures rice produced in India, which it supplements with imports. This is stored and distributed. Other states have their own agencies, but they work with the Food Corporation, and supply it with any spare rice they may have. PDS rice is bought as paddy from farmers at the procurement price, or as a levy of milled rice from millers and traders. The levy varies from state to state, rising according to the marketable surplus in a particular state. In some states the levy is as high as three-quarters of the marketable amount. The levy price includes the procurement price for paddy, plus a fee for milling, handling and transport, and a small profit margin. Despite this, the profit of most millers and traders comes from the sale of the remainder of their rice at current market prices or for export. The government maintains a buffer stock from which general distribution can be made, or any harvest shortfall made up. The PDS draws from these stocks to supply the Fair Price shops, and other subsidised distribution schemes. Those buying from the PDS pay a country-wide standard price, which is fixed according to current market prices and the ability of the poor to pay. If the standard price does not cover the costs of buying and distributing the rice, the difference is borne by the government. Poverty alleviation programmes also use PDS rice, and include schemes to provide employment, to give welfare rice to tribal minorities, and free school meals. A third of Indian rice marketed in India now goes to these schemes, so the PDS actually handles more rice than the entire international rice trade. India however has become an important international trader (Barker *et al*. 1985: 242–6; FAO 1991: 76, 78, paras 22, 24, 26, 27; Roche 1992: 31, 44–8, 75). At the end of 1994 India removed its minimum export price for non-basmati rice, because it had large food stocks, and was expecting another large harvest. Although the government of India wanted to encourage rice exports, the domestic free-market price was above prices on the international market, and the rice was uncompetitive without subsidies which the government refused to give. On 1 September 1994 government stocks were 13.1 million tons, compared with 8.7 million tons a year earlier. So for the international market to draw rice supplies from India, a modest rise in international prices was necessary. Indian rice is welcomed in traditional markets like Saudi Arabia, the United Arab Emirates and other Middle Eastern markets where many Indians now live, and also Iran (*Grain* 1994, November: 18). However 1995 did see a rise in international prices and India emerged as one of Asia's cheapest rice suppliers. In the past, internal price support schemes restricted exports, but rising international prices and shortages of supplies from Vietnam and Myanmar made India competitive. Despite large government stocks, and competitive prices, there were logistical problems in making exports. The major rice export port was Kandla, but this experienced congestion. Rice was taken to Nepal by rail, as it was to Bangladesh where motor trucks were also used. The quality of rice was below usual world trade levels, but sales were made to Nepal, Bangladesh, China, Senegal and even Iran (*Grain* 1995, June: 17). The 1995 exports from India were made from the Food Corporation stock, and were crucial in preventing an international rice crisis triggered by bad weather and resulting bad harvests in China (see Chapter 5). The unexpected emergence of India that year as one of the world's top rice exporters, was due to unusual circumstances. To begin with the government of India lifted restrictions on rice exports. These had been imposed to ensure domestic food supplies at acceptable prices. But government support schemes kept prices high for farmers, and when combined with procurement, storage and transportation costs, the rice from government stocks was over $200 per ton. Up to October 1994 the government kept a minimum export price as high as $225 per ton, which meant that exports would have been unprofitable even if they had been allowed. But by 1995 India had experienced eight consecutive years without a bad monsoon. Rice stocks of over 10 million tons had been running every year since the late 1980s, and in September 1995 they were 14.5 million tons. There was no longer sufficient storage capacity and the government either had to export the rice or waste it by putting it in inadequate temporary storage. India turned to the export market in a big way. Fortuitously this coincided with bad crops in China and Indonesia, forcing them to import rice in very large quantities. At the same time there were low exports from Pakistan, Burma and Vietnam. Even Thailand was in difficulty due to flooding. India was selling in a seller's market! (*Grain* 1995, November: 11–12; *LRBAC* 2 January 1996: 1; see also Chapter 5).

Pakistan

Pakistan has emerged as an important rice exporter since the Second World War, although not on the scale of Thailand and USA. In 1947 the territories which had been British India became independent and Pakistan separated from India. Initially the government controlled the distribution of rice, and held prices down for town dwellers. But low prices for town dwellers also meant low prices for farmers, and discouraged them from expanding production. They could have sold to the world market but an overvalued exchange rate and export controls effectively prevented rice exports. In 1958 a military coup ended constitutional democracy. By 1960 Pakistan was importing large amounts of rice. Then came policy changes and the government increased the price which was paid to the farmers for their crop. The government also began to provide cheap fertiliser and easy credit terms. Moves were also made to return food distribution to private hands, and exports were stimulated by the removal of the export tax. As a result rice production began to rise. This was helped by substantial investment in canal and tube well irrigation. From 1968 production increased in West Pakistan as farmers switched to HYVs. These needed plentiful supplies of fertiliser, which the government provided at subsidised rates. High quality basmati was produced for export to higher income consumers, and lower grades were shipped to East Pakistan. When East Pakistan broke away to become Bangladesh in 1970 this rice was diverted to the world market. During the early 1980s production and exports fell as farmers switched to the lower yielding but more profitable high-priced basmati, and the government withdrew fertiliser subsidies. In the later 1980s basmati sales were hit by the stoppage of sales to Iraq, intermittent sales to Iran, and emerging competition from India. Basmati quality itself suffered as farmers switched to a new HYV variety. The Rice Export Corporation of Pakistan sells all rice from Pakistan (*LRBAC* 30 January 1987: 2). It has experienced difficulties due to bad weather, and the fact that farmers were turning to better earning crops. Basmati quality has also suffered from being mixed with inferior rice. Despite general government control, the private sector was gradually being allowed to export packaged basmati, although the quantities were small (Reynolds 1985: 350–2; Roche 1992: 36, 53, 109–10). The London Rice Brokers' Association reported in 1990 that private traders were now allowed to enter the export trade, but this did not mean much in practice as they could not compete with the prices offered by the Rice Export Corporation (*LRBAC* 31 July 1990: 2).

Bangladesh

Bangladesh broke away from Pakistan in 1970, and since then there have been several coups. Bangladesh is a food deficit country, and has to import grain to survive. Rice is the basis of the agriculture and covers 80 per cent of the cultivated area, although wheat production has been increasing since the early 1970s. Food self-sufficiency has been a target since independence, but population growth has continually outstripped increased production, and the country has had to rely on food aid and commercial imports. The Public Food Distribution System (PFDS) was introduced under the British colonial administration as long ago as 1943 for famine relief. There were two rationing systems, one for the five major towns and the other for the rest of the country. Later schemes differentiated between those who could afford to buy rationed food at both subsidised and market rates, and those who were too poor to buy it at all. From the mid-1970s the very poor were targeted, but many continued to be dependent on rations despite attempts to reduce the number of card holders. Meanwhile the PFDS tried to stabilise market prices at acceptable levels whilst continuing to motivate farmers by buying paddy at prices which would give them a reasonable profit. The third Five-Year Development plan 1985–90 attempted to achieve food self-sufficiency, at the same time providing stable prices at an acceptable level, and alleviating widespread poverty. HYV boro rice and wheat was to be grown during the dry season using irrigation. But floods in 1987 and 1988, a cyclone, and political difficulties, meant that the objective of food self-sufficiency had to be deferred, and imports continued. As a result of the floods the government began discussions with neighbouring countries on flood control. Meanwhile the government continued to buy rice at home at a price which would give the farmers sufficient profit to encourage them to produce. They also maintained control of the trade and distribution of rice, including the keeping of buffer stocks (FAO 1991: 26–7, 28, 30 paras 2, 5–6, 11, 13–16; Reynolds 1985: 371–3).

Myanmar

Burma (since 1989 Myanmar) used to be the largest of the three great rice exporting nations, the rice coming mainly from the Irrawaddy delta. But after the Second World War and independence from Britain, exports went into severe decline. Independence came in 1948, and with it a desire to isolate themselves from the world economy. Although the land was nationalised, tenants were given effective possession. The Indian population fled the country, and their land was redistributed to Burmese nationals. A limit was put on the size of individual holdings, so between 1948 and 1965 the number of farms doubled, although most were only between two and four hectares. But the government provided little in the way of irrigation, or even fertiliser and new varieties. Rice exports, previously so important, were neglected. Government marketing boards were established to buy and distribute rice, offering local farmers prices well below the world level to ensure cheap rice for town dwellers. These low prices, coupled with the disorganisation of the marketing boards and the distance from the farms to the buying centres, discouraged farmers. So they turned to the black market, or grew other commercial crops which they could sell for a better price, or simply provided for their own use. Denied the incentive of profit, few were prepared to invest in improving their farms. Surprisingly, paddy production had recovered to pre-war levels by 1962, but that year there was a military coup and a long period of military rule. Rice output now ceased to grow, even though the population continued to increase. The exportable surplus declined, and exports fell to about a tenth of what they had been before the war. Privately owned mills and storage facilities fell into disrepair as owners feared they would be taken over by the state. Overvalued exchange rates hit exports by making Burma rice overpriced in world markets. But in 1972–3 after a poor harvest and a rice shortage, government purchase prices were raised to encourage farmers, and high-yielding varieties and subsidised fertiliser were made available. The purchase price was raised again the following year, and domestic production recovered. Burma was initially slow to adopt the new HYVs as consumers did not enjoy the taste of them. Also the semi-dwarf height of the early varieties was unsuited to Burma because most rice was grown using natural flooding, and the water levels varied too greatly. In 1974 there was a change in government, and after many years of isolationism discussions began with the World Bank about obtaining loans with which to bring about economic reforms. As a result the 'Whole Township Programme' was introduced which brought in subsidised fertiliser and HYVs. Paddy prices were raised, and by 1975 were twice what they had been in 1970. Sales to the free market were again allowed. As a result yields and output rose, and so exports began to recover. But in the 80s production again fell away because government ceased to increase prices, even though there was inflation which hit farmer's input costs. Fertiliser use fell because it could not be imported due to a shortage of foreign exchange. Rice exports were an important source of foreign earnings, but without rice exports, the fertiliser to grow the rice could not be imported! Although by the late 1980s HYVs were produced on three-quarters of the cultivated land, the farm area had decreased, and milling yields from the old run-down privately owned mills were declining. So from 1987 the government ended its policy of compulsory rice purchase of about a third of the crop at low price levels, and legalised inter-provincial private trade. Rice prices rose, and government purchases fell. So measures had to be re-introduced to obtain rice for the army, civil servants, and for export. Despite fertiliser shortages, higher prices resulted in higher rice production. Yet exports declined because rice was smuggled to neighbouring countries where better prices could be obtained. This was largely due to the fact that the official exchange rates was so high that Myanmar rice could not compete in world markets. From 1987–8 private exports of rice were allowed, but very high exchange rates continued to deter foreign buyers. As government purchases fell, rice hoarding began, and prices rose. This caused rioting and warehouses were robbed. Exports ceased. Now that rice could be moved between provinces, small village mills were set up, and there was increased smuggling to Bangladesh, with some also going to China and Thailand. So Myanmar exports continue to be low, and deals are usually done on a government-to-government basis, the sales being made by Myanmar Export and Import Services. There are however some sales to international trading companies, and occasional tenders for job lots usually of brokens. Mauritius, Sri Lanka, China and Bangladesh have been markets for Myanmar rice, and the quality is usually low quality 25 per cent brokens and brokens (Reynolds 1985: 146–9; Roche 1992: 30–1, 42–4, 117). In 1992–3 the Ministry of Agriculture and Irrigation tried to increase the area and output of the summer or second rice crop, planted between November and December and harvested March to May. This was to try to expand the surplus available for export, in order to improve hard currency earnings. In December 1995 the government was still hoping to achieve this increase. That year they also promoted double cropping of the main crop monsoon rice, planted in June to August and harvested between October and December. But these plans were impeded by Myanmar's lack of irrigation, shortage of equipment for threshing and drying, and lack of fertiliser and insecticide. In any case, the farmers preferred to grow pulses, the normal second crop, which are better suited to the dry season, and more profitable than rice. By 1997 the Ministry of Agriculture and Irrigation's plans that there would be 4 million acres of second crop rice yielding 5 million tons of paddy were scaled back by a half to 2 million acres. The second crop was not producing a surplus for export, and was making the domestic supply situation unstable. So the Ministry changed its strategy, and decided to help farmers produce pulses, cotton and sugar as second crops for export. Farmers were to be allowed to decide what to plant according to their understanding of the market and the farming conditions in their area. Myanmar exported 619,000 tons of rice in 1994 and 645,000 tons in 1995, but in 1996 exports only reached 265,000 tons. It is thought that the high export levels of 1994 and 1995 were only secured by reducing stock levels, and not because the crop surplus grown for export had risen. The priority of the government was to grow sufficient rice for domestic needs and stabilise prices, and to export only when there was a surplus. However the government still wanted to obtain foreign currency earnings, and it seems that they hoped to obtain these by encouraging farmers to grow the premier grades suitable for export (*Grain* 1997 March: 11–13).

Malaysia

In the nineteenth century Malaysia used to import rice from Burma, Thailand and French Indo-China. But the depression of the 1930s hit export earnings from tin and rubber, and the lack of foreign currency meant that the colonial government encouraged local rice production so that imports would not be needed. Cheap land, irrigation and seed was provided. With the outbreak of the Second World War, rice prices were regulated and minimum prices offered in some states for rice delivered to government mills. But even in the early 1950s about 40 per cent of rice consumption was provided by imports. With independence from Britain in 1957 rice self-sufficiency was targeted, and to achieve this there was heavy investment in irrigation. The World Bank helped finance the Muda River Irrigation System, and the construction of irrigation networks which would support double cropping were given priority. Paddy production doubled in twenty years. New varieties and increased fertiliser application all played their part. Irrigation expansion and fertiliser subsidies continued during the 1970s, and there was price support so that farmers continued to have an incentive to produce. The government aim was to keep imports to 10–20 per cent of requirements. Rice remains the principal cereal, but during the 1980s the area under rice declined steadily, as commercial crops like coconuts and palm oil gave farmers better profits. So import requirements have increased to about a third of total consumption, or 0.5 million tonnes out of 1.5 million tons. This dependency on imports seems destined to increase unless greater encouragement is given to local farmers (Barker *et al.* 1985: 252–3; Latham 1981: 164; 1983: 260–80; Roche 1992: 79, 131; Wickizer and Bennett 1941: 165–87).

Indonesia

In the late nineteenth century and early twentieth century, the Dutch East Indies or Netherlands India, now Indonesia, exported items such as sugar, rubber, tin and petroleum. With export earnings they purchased rice from Thailand and French Indo-China. But the coming of the depression after 1929 meant exports declined, leaving a shortage of foreign exchange to buy rice from abroad. This forced the Dutch colonial authorities to adopt a rice self-sufficiency policy. When the Second World War broke out, a commission was established to buy and sell rice in order to maintain prices and ensure adequate supplies. When Indonesia became independent from the Netherlands in 1945 the government tried to continue these policies so as to ensure cheap supplies to urban areas. But the size of the civil service was increased, and rice rations were supplied to civil servants and army personnel in 1951 and 1952 as part of their wages. This was to ensure their loyalty at a time of soaring inflation. Substantial rice imports were made in these difficult years to fulfil these demands, but gradually rice production increased and by 1954 had become more stable, and it was possible to reduce the inflow. The increased availability of rice made it possible to cease the rice rations to public servants and the army in 1953 and 1954. However rice production fell in 1955 and 1956 and substantial imports had to be made again. This shortfall in rice provision was accompanied by inflation caused by government overspending. So once again rice rations had to be supplied to public servants and the military. Provincial governors were required to obtain rice for the government, and inter-provincial trade was banned. Provinces which were short of rice saw prices surging, whilst provinces with plenty of rice found prices falling. This meant that rice was available for the government from provinces with surpluses more cheaply than if market forces had been allowed to operate, which is exactly what was intended. However, as elsewhere in Asia where such policies were applied, the system discouraged farmers from expanding production in the provinces capable of providing a surplus. At the same time the cost of importing rice depleted the foreign reserves. The shortage of foreign exchange restricted imports, and farmers were required to sell rice at low and unprofitable prices, which made the situation even worse for the farmers. So in 1959 the Sukarno government turned to a new agricultural scheme based on village centres. At the village depots seed and fertiliser was made available, together with advice to farmers on their use. Loans were also available, to be paid off with paddy. But the scheme proved ineffective, and crops declined from 1959 to 1963. Imports were high again in 1964 when they rose to a peak of nearly 2 million metric tons and inflation raged in 1965. That year an attempted communist coup against the Sukarno government was suppressed by the army, which then established a military government under General Suharto in 1966. The new government inherited a bankrupt treasury, empty state rice warehouses and raging inflation. The worthless old currency was withdrawn, and a new rupiah introduced, worth 1,000 old rupiahs. By controlling the supply of the rupiah, inflation was brought under control. Yet rice prices tripled in 1966, and the situation worsened in 1967 due to drought which affected the dry season crop. There was still a shortage of home produced rice, making imports essential, and causing prices to fluctuate. The government moved to control rice and fertiliser prices at the current level. Since the early 1960s Bogor Agricultural University had run a programme called DEMAS (Mass Demonstration) to improve rice production by showing farmers how to use modern seeds, fertiliser and insecticides. Yields had risen by a half in areas where the programme operated. Now the Suharto government took over the scheme and extended it widely, under the name BIMAS (Mass Guidance). Cheap loans were given to farmers, so they could obtain better seed and fertiliser. The purchase of their crop was guaranteed, and improved management of water was provided. But supervision of so many farmers proved costly, and a new scheme was brought in called INMAS (Mass Intensification) to get farmers to work independently, and pay for their own fertiliser and insecticides without loans. Rice production increased, but because of a shortage of foreign exchange, in 1968 the government asked foreign manufacturers to help provide fertiliser, insecticide and advice. BIMAS GOTONG ROYONG (BIMAS Mass Help) only lasted for a couple of years because individual companies specialised in insecticides for particular pests. To target specific infestations the manufacturers preferred cheap mass application techniques over wide areas, like aerial spraying, which they could do whether the farmers wanted it or not! They did not want to deal with the small scale pest control problems of individual farmers. This was also the year that the first semi-dwarf HYVs were brought to Indonesia, and these were soon bred with native rices to produce pelita, a rice closer to the tastes of local consumers. These new rices were used in the BIMAS and INMAS programmes. The army and civil servants continued to receive rice rations, but lack of foreign currency, and a world shortage made it impossible for the government to import enough rice to keep prices down. Prices tripled yet again. So from 1969 the government raised paddy prices to stimulate production. The BIMAS GOTONG ROYONG programme was abolished in 1970 because so many farmers had defaulted on their loans, and a new BIMAS scheme was introduced giving farmers credit for packages of inputs suitable for their particular farm. National basic prices for paddy and fertiliser were also introduced in 1970, which made it worthwhile for farmers to increase production. There were also fertiliser subsidies. The government began work to restore the irrigation system which had been neglected under Sukarno, and more HYVs were introduced, increasing double cropping. Most of the irrigated area was on Java, but schemes were introduced to expand irrigated areas in Sumatra and elsewhere. The new BIMAS scheme was optional, but inputs were delivered quickly when needed, and better prices were paid for the resulting paddy. The village centres were invigorated by being allocated a representative from the government bank, an adviser, a fertiliser supplier, and two local assistants. They monitored the farm performance of the district. With good weather the situation improved but 1972 was another bad year affected by drought, at a time when rice producers and consumers in neighbouring Thailand, China and the Philippines were also hit. Emergency imports were made as rice prices again soared, and rice had to be imported yet again in 1973. This was the year of the great oil crisis, and the increase in OPEC oil prices which followed in 1974 meant that the government made greater foreign earnings from oil sales. Food security and stability in rice prices was the responsibility of the National Logistics Agency (Badan Urusan Logistic or BULOG), which reported directly to the cabinet. It had a network at provincial level and district levels which linked with farmers, traders and village co-operatives. As a result of the improved oil earnings, more foreign currency was now available for rice imports. BULOG was able to bring in foreign rice to keep prices down, and also build up a buffer stock to bolster national supplies in case there was a sudden crop failure. At the same time, better paddy prices were paid to encourage the farmers, and loans were made to them to buy improved seed and fertiliser. Indonesia also received 2 million tonnes of rice as food aid between 1973 and 1980. Yet domestic production stagnated even though incomes were rising, and so was consumption. Dependence on imports grew worse. By the late 1970s Indonesia was the world's largest rice importer, taking about 20 per cent of world trade. BULOG skilfully spread its purchases widely, so as not to trigger off a general increase of world prices. It also made government-to-government contracts tied to oil, and bought from particular favoured countries in good harvest years, when the rice could be obtained easily elsewhere. The aim was to build up a network of regular suppliers who would return the favour in times of shortage. BULOG also bought from a limited number of international rice traders who could be relied on not to disclose the source of their orders. One of BULOG'S aims was to secure a regular year round flow of rice so there would never be a sudden interruption in supplies, even if their own harvests were poor. Indonesia wanted to avoid the situation in which it was obvious that they had a shortage as this would mean that world prices would be raised against them. To try and reduce the growing import dependency, assistance schemes to farmers were improved. In 1979 INSUS special intensification schemes were introduced, and improved BIMAS programmes, which put between 20 and 50 farmers into groups, and co-ordinated their planting, application of insecticides and irrigation. The group was also responsible if individual farmers did not repay their production loans. So by 1981 rice production had increased, but not sufficiently for Indonesia to manage without imports, and these went on increasing. In the 1980s Indonesia finally attained rice self-sufficiency. In 1986 came the SUPRA INSUS project, augmenting BIMAS and INSUS. Later there was also OPSUS (Cultivation Extension). The fertiliser subsidy was reduced, and the rate of growth of production began to slow as the spread of HYVs reached its limits. Irrigation investment was curtailed, and pesticide use was restricted. Because Indonesia had achieved virtual self-sufficiency rice imports fell, and tended to be only of the highest qualities. Between 1985 and 1987 Indonesia was a net exporter of rice, although costs of production made the rice uncompetitive in world terms. Bad weather and falling domestic production forced large imports again in 1989. BULOG continues to control price stability and ensures food security by holding stocks. There is a minimum price, stated by the Economic Council of Ministers well before harvest time to encourage farmers. There is also a maximum price so that consumers are protected. BULOG maintains this price spread by buying from farmers and selling to wholesale markets. BULOG is also responsible for providing the rice allocated to civil servants and the army, and has the authority to import rice if it is necessary (Barker *et al*. 1985: 249–51; FAO 1991: 92 paras 9, 10; Latham 1981: 165; Roche 1992: 32–4, 48–9; Wickizer and Bennett 1941: 165–87). As noted above Indonesia was the world's largest rice importer in the late 1970s, and in the 1990s has been the third largest consumer of rice in the world. Their trade position is an important factor in the world rice market, maintaining a policy of 'self-sufficiency on trend' by which the aim is at self-sufficiency, but if there is surplus then they export. If there is a deficit, they import. Production was only about 24 million tons of paddy in the late 1970s but averaged almost 46 million tons in the early 1990s. But 1994 saw a shortfall due to drought conditions, and Indonesia purchased large amounts of low quality rice from Vietnam and Thailand (*LRBAC* 31 July 1994: 1, 3; 30 September: 1). There was a further shortfall in 1995, and more imports from Thailand and Vietnam. BULOG manages the rice stocks, and makes all imports and exports. When exports are made, BULOG prefers to make them in the form of rice 'loans' which are to be repaid in the future in rice when it is needed. This way BULOG can offload old surplus rice, and avoid warehousing costs, but receive equivalent amounts of fresh rice back when there is a shortage. A good example of this was in 1994 when BULOG asked the Philippines to repay by 31 December loans of about 200,000 tons made before 1990. There was to be no extension of the time limit. The loan could be repaid in cash or kind, and carried an annual interest rate of 6 per cent. As the Philippines themselves had a rice shortage, the rice had to be obtained from a third country. BULOG manages the domestic crop by buying rice for stockpile when there is a surplus and releasing rice from the stockpile when there is a shortage. So it is not always easy to understand what the actual situation is at any one time. But in 1994 and 1995 Indonesia bought so much low quality rice from Thailand, Vietnam and Pakistan that they forced up world prices. Lesser amounts came from Myanmar. The United States was also a source of supplies (*Grain* 1994, July :10; September: 14–15).

Thailand

Thailand has been an important rice exporting country since the late nineteenth century with most of its supplies coming from the Chao Phraya delta. Before the Second World War it was less important than Burma, and about on a par with French Indo-China. But whilst political and economic difficulties removed these other countries from the international market after the war, Thailand continued to be a major supplier. Since the war the United States of America has been their main rival. Formerly the Thai rice trade was in private hands, but after the war the government introduced an export surcharge, known as the 'rice premium' and rice export quotas (see Tsujii 1977a: 202–20). Whilst this provided revenue for the government and kept rice prices down in Thailand, it also meant lower paddy prices for farmers.

Although they received low harvest prices, their fertiliser costs were higher than in their Asian rivals many of whom received subsidised fertiliser. Despite this high cost, low price situation, the area under rice expanded. The government expanded the area which was irrigated, and also the rural road network so that rice could be transported more easily. Between 1964 and 1971 there were major irrigation schemes with the Bhumipo and Sirikit reservoirs, and from 1969 HYVs were introduced. But fertiliser imports were restricted in an attempt to encourage local producers to supply the country's needs. After the oil crisis of 1973 oil prices soared putting hard currency in the hands of major rice consumers in the Middle East, West Africa, and Indonesia. This led to a huge increase in rice exports, although it was not until the mid-1970s that the export tax was reduced and fertiliser prices cut. From 1976 large government-to-government sales were made to Indonesia and Malaysia. In the early 1980s Thailand increased its share of world exports at the expense of the US because their prices were lower. This continued even when the US provided improved loan rates and subsidies to buyers. There were further irrigation improvements in this period, including private tube wells. The water was used for dry season HYVs, which were exported as these rices did not suit Thai tastes. Traditional methods are still used for the main crop, and fertiliser inputs continue to be low. So yield is only 2.0 tonnes per hectare which compares badly with rival countries using modern methods. In 1986 the export premium on low quality rice was abolished, except for a small local tax, and the quota system was dropped (*LRBAC* 31 January 1986: 1, 3). In December 1987 a serious situation arose in

Bangkok and there were many defaults by exporters. Prices rose sharply at the end of December as the new crop was harvested and merchants realised how small the crop had turned out to be. Local merchants tried to buy up all the available paddy, and then would not supply it to the mills. Thai exporters who normally found that prices fell at harvest, now found they had to pay very high prices to get rice to fulfil contracts they had already agreed to, or that they could not get any rice at all. In consequence many defaulted, causing a serious loss of confidence by foreign buyers. As previously noted in Chapter 5, the *London Rice Brokers' Association Circular* observed 'If exporters expect to default when the market goes up, how can they expect their buyers not to default if the market goes down?' (*LRBAC* 29 January 1988: 1, 3). In 1989 rice was Thailand's third largest export earner after tourism and textiles. That year there were huge sales to Indonesia and China. In the following years China took less rice leading to a decline in government-to-government sales. This was accentuated when Malaysia, a regular customer, began to buy from the private sector. Export policy is decided by the Department of Foreign Trade of the Ministry of Commerce. The Rice Committee of the Board of Trade meets every week to decide export prices, which includes selected members of the Thai Rice Exporters' Association (see Chapter 5). Other questions of rice policy are also dealt with by the Committee, which advises the government, and the Rice Inspection Committee decides the grades of rice. Whenever prices have fallen, government policy has been to buy milled rice at above the going rate in order to sustain the price of paddy. The sellers then stock the rice, and receive a down payment of 20 per cent from the Ministry of Commerce. The rice can then be used as collateral for cheap loans from the Bank of Thailand. The Ministry then pays the Bank of Thailand the difference between the low interest loan rate and the commercial loan rate. The stocks of rice are then sold in government-to-government deals. So the government has encouraged farmers by supporting high paddy prices, and at the same time has provided subsidies to maintain competitive export prices. But it is moving to the view that the market alone should determine prices. More than two-fifths of Thai rice is usually consumed in Asia, with Africa taking a fifth, and the Middle East another fifth. Asia takes high quality 100B rice, particularly Malaysia, Hong Kong and Singapore, and Iran is another major purchaser of this grade. Africa takes lower qualities with a higher percentage of brokens (Barker *et al.* 1985: 251–2; Reynolds 1985: 162–3; Roche 1992: 36–7, 55, 102–8). In 1994 the Ministry of Commerce and the Thai Exporters'

Association met to implement a rice export subsidy programme similar to that already operating. The scheme was for forward sales of 10,000 tons or more of medium and low quality rice (10 per cent or greater broken). For each sale, exporters received a fee of $10 per ton, up from $5 per ton in 1993. Recent export support programmes have subsidised up to 40 per cent of Thai exports. Countries to benefit 1994–5 were China, Indonesia, Bangladesh, Senegal and various other African countries (*Grain* 1995 January: 12–13; July: 10–11).

Vietnam

French Indo-China which included the Mekong delta was another of the three great rice exporting areas before the Second World War, but like Burma, it too ran into difficulties subsequently. Firstly there was the war against the French, and then civil war in Vietnam between the North and the South which embroiled the USA. As the Vietnam war took hold, rice farming was affected, and in the 1960s exports ceased and imports began. Harvests declined between 1965 and 1969 and in the early 1970s imports grew. Both the North and the South imported rice, the North from China and the South from the USA. There was some recovery in domestic production in the later years of the war up to the cessation of hostilities in 1975. This was helped by land reform in the South, the expansion of the irrigation system, and the introduction HYVs (see Tsujii 1977b: 263–94). At the end of the war both China and the USA stopped supplying rice. But the new government in the South imposed collectivisation which farmers opposed. The distribution of fertiliser and other inputs was centralised, but handled inefficiently and not enough was supplied. There was also a shortage of bullocks needed for pulling ploughs and wagons. So in 1981 the contract system was introduced by which farmers were allowed to run their farms themselves, and keep for themselves any rice produced over and above the allocation contracted by the government. This could be for home use or sold to the private trade. The system lasted until 1987, when bad weather caused problems and there were shortages of fertiliser and other inputs. The problem with the contract system was that farmers were not free to decide what crops to grow. The return to the farmers was often low because the government did not always buy the amount of rice it had contracted for. Farmers were also unwilling to make any investment in their farms because they no longer owned them. 1987 saw a bad harvest, and to try to give added incentives to farmers, land was assigned to families on leases of 10–20 years, so that the farmer and his children would enjoy the benefit of any investment in land improvement. As from the beginning of 1989 the contract system was abolished, and replaced by a simple land tax paid in grain or cash. The contract rice had been used for cheap rice rations for the army and civil servants so these handouts were also abolished. At the same time private trade in rice was legalised. Control of fertiliser and input supplies was transferred from central to provincial authorities, and while the import of these items remained in state hands, private traders were allowed to sell them to farmers. As a result rice production rose and in 1989 Vietnam again returned to the world market as a major exporter (see *LRBAC* 29 December 1989: 1, 3). Despite the problems of the 1980s the irrigated area had increased, and so had fertiliser use, and the cultivation of HYVs, all three being linked (see Chapter 2). In the 1990s the government has made the decision about the minimum export price (MEP), setting it below their chief rival Thailand, but even these prices are undercut illegally, and considerable smuggling takes place especially to China. Vietnam now charges feed grain prices for low grade rice, and has sold to India, Sri Lanka, China and Latin America. But higher qualities have gone to USSR and Iran, in place of Thai rice. Vietnam has now become once again one of the great rice exporters, after Thailand and the US, much as they were before the Second World War. Private traders are allowed to flourish, and international brokers have set up in the main markets. Farmers tenure of plots has been extended to 50 years, farm taxes have been lowered, and farmers are no longer required to sell to the state and can sell direct to the market. New Japanese rice mills and equipment have been obtained to upgrade quality. However there are still doubts about the quality of Vietnam rice, and their capacity to complete orders (Roche 1992: 37–8, 55–6, 110). Late in 1995 the government of Vietnam announced that it was to reorganise its rice export sector. Although Vietnam was again one of the world's top rice exporters, dozens of firms sold rice abroad, earning important hard currency for provincial and local governments. This was thought to result in lower export prices as the firms competed with each other for sales. This reduced earnings for the state owned export firms Vinafood I and Vinafood II. So plans were made to concentrate rice exports in the hands of the new Southern Food Corporation (SFC) based on the leading rice exporter Vinafood II, based in Ho Chi Minh City. Hanoi based Vinafood I was to continue small amounts of exports on government-to-government contracts. Under the export quotas for 1996 only fourteen companies were authorised to contract for new exports. These were controlled either by the central government, or provincial governments (*Grain* 1996, February: 10–11). However, in January 1997 it was reported by the *Financial Times* that Vietnam was to relax the state's monopoly on rice trade by allowing private companies to export rice. Multilateral institutions have pressed for many years for Vietnam to liberalise the rice trade to introduce competition and stimulate more exports. The Ministry of Agriculture reported that despite widespread flooding and storm damage, Vietnam exported a record 3.05 million metric tonnes in 1996, as against 2 million metric tonnes in 1995. This was because in October the government increased the export quota from 2.5 million tonnes to 3 million tonnes. The decision on private exporters was signed by the trade minister Mr Le Van Triet, but it was not indicated how this would link with the quota. The Ministry of Agriculture said that Hanoi might have to adjust the export quota, as reports from distant provinces indicated that the bad weather had destroyed rice stocks, and people were starving. There had also been flood damage in the last six months of 1996, making it difficult to predict the export volume in 1997. IRRI estimated that Vietnam was the world's second largest exporter in 1996 (*Financial Times* 8 January 1997: 24). USDA *Grain: World Markets and Trade* published a feature on Vietnam in their February 1997 edition, subsequently published in *Rice Journal* March 1997. Morgan Perkins, the author, says that Vietnam began to emerge as a major rice exporter after the *Doi Moi* policy reforms of the Vietnam Communist Party in 1986. These guaranteed land tenure to farming families, and opened the rice market to private dealers. In the previous ten years Vietnam had been a major rice importer, but in consequence of the reforms rice production had increased by two thirds and they had also become a major exporter. Even in 1988 Vietnam was a net rice importer, but in 1989 they became an exporter, with overseas sales of 1.4 million tons out of a world trade of 14 million tons. Exports have since grown steadily, and in 1996 were 3 million tons or 16 per cent of world trade. To begin with Vietnam only exported low quality rice, and so presented no threat to Thailand and the United States. But in the mid-1990s she began to sell high quality rice to Iran, Singapore and Malaysia, consumers of the best qualities. Rice exports now constitute 20–5 per cent of Vietnam's export revenue. Exports are largely by state trading companies, which are heavily influenced by the ministries of Trade, Agriculture and Rural Development, and Planning and Investment. The Northern Food Corporation and the Southern Food Corporation make requests for export quotas which are reviewed by the above ministries, and the quotas are then issued by the Ministry of Trade. Usually there is an initial quota followed by an additional quota when the late harvest is in. When the contracts are ready to be executed, exporters apply again for an export licence, granted on a shipment-by-shipment basis. Export licences are often refused if the contract price no longer matches actual market prices. This has caused problems for foreign buyers of Vietnam rice, as there are great and sudden fluctuations in world rice prices, and implementing contracts in Vietnam can be slow. The two major export firms are linked to the government, and Southern Food Corporation (created from Vinafood II) in Ho Chi Minh City is the largest exporter, with up to 25 per cent or more of exports. SFC can also influence the share of quota received by other exporters. SFC handles the large government-to-government contracts with regular purchases such as the Philippines, Cuba and Indonesia. SFC also effectively sets the current price, and other exporters keep their prices close to those of SFC. The Northern Food Corporation, which also goes under the name Vinafood I, is located in Hanoi, and specialises in government-to-government contracts with politically sensitive allies. In 1996 most of Vinafood I's shipments went to Iraq, although these were under 200,000 tons. There are other exports made by food companies linked to provincial governments. Most of these are in the Mekong delta, where the majority of the rice exports come from. Because these exports give the provincial governments funds for imports and business initiatives, they have resisted attempts to centralise rice exports in the hands of SFC and NFC. Foreign rice firms have been establishing themselves in Vietnam, and because they cannot legally export rice, they act as agents for provincial food companies, taking a 1–2 per cent commission of contract value. Their marketing expertise has been crucial in re-establishing Vietnam as a major exporter. Foreign companies have also entered into joint ventures in milling and processing. American Rice, Inc. operates a mill in the Mekong delta, and the Golden Resource Corporation has a processing plant in Ho Chi Minh City. Both operate in conjunction with the Southern Food Corporation. These companies have been crucial in making high quality rice available for export, although American Rice in particular has been given much smaller export quotas than were agreed when the joint venture was set up. What is even more galling is that even though the foreign partners put up most of the money for improved equipment, the local partners have been unwilling to allow them to make the decisions as to what plant should be purchased. In a recent report the Asian Development Bank was very critical of the way Vietnam's rice export sector is operated. It seems unlikely that despite new regulations, there will be much immediate change. No significant new rice varieties are likely to give Vietnam big increases in production in the near future, and irrigation expansion is slowing as marginal costs are rising steeply. Many farmers are also returning to low-yielding high-quality rice varieties. Vietnam's future in the international rice market depends upon improving the reliability of contracts and export certification, and allowing foreign firms to bring in their expertise. *Rice Journal* adds a comment that Vietnam will only become a major competitor of the United States when their processing capacity is improved and increased, including rice storage and drying facilities, and better milling technology (*Grain* 1997, February: 7–10; *Rice Journal* 15 March 1997 : 24–8). The *Financial Times* reported on 2 April 1997 that Vietnam was to decentralise the allocation of rice export quotas, and abolish barriers restricting internal trade. In 1996 Vietnam was second only to Thailand as a world exporter, earning $1 billion in foreign exchange. But the new rules, contained in two government decrees, fall short of allowing private sector participation in rice exports. Under the decrees Hanoi set a rice export quota for March to September of 2.5 million tonnes. Previously the quota was allocated to Vinafood I and Vinafood II, who took the majority, spreading the remainder between thirteen other state companies with export licences. In this way the two major companies could take advantage of differences between local and world prices, to the disadvantage of local farmers. But under the new regime, two-thirds of the quota will go to the fourteen provinces of the Mekong delta, where most of the rice comes from. They can then nominate a company to handle the exports, of which there are twenty-one including both Vinafood companies. These new measures represent a victory for Hanoi over the entrenched interests of the two Vinafood companies. 'The next steps towards full liberalisation are likely to be pushed by the provinces themselves' according to Mr Francesco Goletti of the Washington International Rice Research Institute. The decrees also abolish taxes and licensing procedures which have restricted rice trading between northern and southern provinces. This should stimulate the domestic rice trade, and allow prices to be set by market forces rather than the Vinafood companies as previously. Domestic rice traders will be allowed to operate. Mr Nguyen Dang Chi of the trade ministry says traders must secure a licence or business certificate, have a minimum capital of 50 billion dong ($4.5 million) and have been trading for three years. However very few local private companies are large enough to meet these criteria! (*Financial Times* 2 April 1997: 31).

COUNTRIES AND POLICIES

The Philippines

The Philippines has long been a rice deficit country. Before the Second World War imports came mainly from Saigon-Cholon in French Indo-China, now Vietnam. As a response to the depression, rice control was introduced in the Philippines in 1936, the government buying paddy, milling it and distributing it. This was done deliberately to exclude the Chinese middlemen and millers who had previously profited from these tasks. As a result the government controlled the market and prices to both farmers and consumers. In the early years of the century imports supplied 5–10 per cent of consumption. After independence in 1946 the aim of government policy was to keep prices below agreed levels, and avoid fluctuations. To do this imports had to be made, and these were sometimes very large, as in 1951, 1958 and 1963–7. From 1965 a self-sufficiency policy was adopted, made possible by the coming of HYVs and the so called 'Green Revolution'. The Philippines felt the impact of the new methods before other Asian countries because the International Rice Research Institute (IRRI) was located at Los Baños not far from Manila, and the new techniques were applied there first. Self-sufficiency was obtained in 1968–70 but in 1971 substantial imports had to be made to stabilise prices at an acceptable level. There was another bad year in 1972 and in 1973, the year of the oil crisis, rice was virtually unobtainable on the world market. So there were greater efforts to achieve self-sufficiency, and irrigation investment was increased. In response to the crisis of 1973–4 the government distributed a mixture of rice and maize to keep prices down to the agreed level. When the crisis was over, the government introduced a scheme called Masagana 99 which combined low interest credit, subsidised fertiliser and the new methods. There were substantial subsidies for fertiliser to begin with, but after 1976 they were reduced. This scheme was accompanied by continued investment in irrigation, which probably had more effect than anything else in increasing Philippine rice production in this period. The fact that production continued to increase even when loan interest and fertiliser subsidies were reduced, suggests that irrigation expansion was indeed the key factor. The government fixed a bottom price for paddy for many years, but market prices were usually above this. When prices fell below the minimum rate, the government did not have the money or administrative structure to buy from the farmers. With the Masagana scheme minimum prices were more effectively maintained. Rice production continued to expand during the late

COUNTRIES AND POLICIES

1970s, helped by irrigation expansion, good weather and shorter maturing varieties. As the crop surplus grew, the government increased storage capacity, and the possibility of exporting rice was considered. Yet despite achieving rice self-sufficiency, grain imports of rice, wheat and maize continued to grow. The Philippines was one of the few countries in east Asia to experience rapid and sustained growth during the 1970s, but in the early 1980s there was a severe downturn with real declines in GDP in 1984 and 1985. The growth of rice yields and output also slowed in the 1980s, and there were actual falls in the area cultivated. But per capita income remained well above most Asian countries including Bangladesh, China, India and Indonesia. Rice continued to be the most important cereal, followed by maize, rice providing about 20 per cent of total food consumption, and 40 per cent of calorie consumption. Grain imports have continued, although recently more wheat has been imported than rice, and this seems likely to continue. One feature of the Philippines is that its per capita consumption of rice is likely to fall in the future, because average incomes are at the crucial level where meat, fish, vegetables and fruit are substituted for rice (Barker *et al.* 1985: 253–6; Latham 1981: 166; FAO 1991: 114–16, 124 paras 1, 6, 10–11, 25–6).

China

Before the Second World War, China was a major rice importer, with rice coming from Thailand and French Indo-China by way of Hong Kong (Latham and Neal 1983: 260–80). Both rice and noodles are eaten in many parts of China, and as wheat is cheaper than rice, noodles can be substituted when rice crops fail. Output of paddy in China is usually double that of wheat, and about 30 per cent of the grain yielding land is under rice. Rice is the main crop south of the Yangtze river, and to the north, wheat and other cereals dominate. The central and eastern regions are the most important rice producing areas, followed by the southern and south-western regions. The north and north-east produce very little. About 60 per cent of the agricultural workforce is involved in producing rice, and China produces and consumes about a third of the world's total. During the twentieth century rice has moved north and wheat south where circumstances have permitted. Maize is another important crop. When the People's Republic of China was established in 1949 China had no proper agricultural research system. Land reform was carried out in 1951 and 1952, land being confiscated from major landowners and divided into small private holdings. There were no artificial fertiliser plants, and the transport system was so bad fertiliser could not even be distributed to farmers or supplied to them at acceptable prices. Agricultural development depended on water control and early maturing varieties, and between 1952 and 1957 there was an increase in multiple cropping. But agriculture was given a low priority in China's economic strategy, and had to supply resources to the industrialisation programme. The public food distribution system was introduced in 1953 following the implementation of a state monopoly system on grain purchase and distribution. Production units were classified as grain surplus, self-sufficient or deficit and a fixed quota for production, procurement and consumption was set based on a 'three quota' ruling. Production teams sold surplus grains to the State Foodgrain and Edible Oil Bureau which was responsible for their distribution to consumers. In 1958 came collectivisation and the creation of communes. But collectivisation and the loss of their own farms resulted in farm workers losing the incentive to produce, and coupled with bad weather, there was a disastrous fall in production. Management decisions and the sharing of profits had to be taken from the central committee and given to the production teams. These were usually groups of about thirty families farming 10 hectares of land. Until 1965 the government purchased rice at fixed 'quota' prices. In 1965 a double price system for buying paddy was introduced. This secured low prices for consumers. Each production unit had a quota and the quota rice was bought at a price well below world price levels. Any surplus rice produced above the quota was bought at a higher price than that given for the quota, to encourage the workers to produce extra rice. Marketing and the distribution of rice was subsidised to provide low prices for consumers and there was rationing to hold demand in check and achieve fair distribution. There were attempts to reverse all this during the Cultural Revolution which began in 1966, a period which also led to interruption of agricultural research and training in modern methods. Despite its failures in incentive terms collectivisation did mean labour could be allocated for irrigation schemes and manure production, and targets could be monitored. Labour used for manure production accounted for nearly a third of total labour input. Despite the problems of this period modern HYVs were introduced, and China developed its own semi-dwarf varieties and F1 hybrids. The first of the new semi-dwarfs were introduced in 1964, two years before the release of the first IRRI variety. In the autumn of 1970 a male-sterile wild rice was found on Hainan Island, which led to the breakthrough in the breeding of hybrid rice. In 1971 a hybrid rice breeding programme was started in Hunan Province. By 1974 the first demonstrations of hybrid rices were successful, and these were then introduced widely. Semi-dwarfs were grown on 80 per cent of all rice acreage by 1977. Yet despite these innovations rice production stagnated in much of China. The price for rice produced surplus to the quota was raised in 1972, but as it was then held at the new level until 1978 the incentive effect was soon lost. Meanwhile double cropping and even triple cropping was being established for the main quota crop. After Mao's death in 1976 there were substantial changes in agricultural organisation.

In 1979 the 'responsibility system' was introduced, and although the land remained under common ownership, the state contracted on short leases with individual farmers and their families for a share of the harvest, or its equivalent in cash. Each household was given a target quota to achieve, and the government continued to control the domestic rice trade. But as previously, farmers could sell to the government any rice they produced surplus to quota at a better price than they got for their quota. In return for their quota rice they were given subsidised diesel fuel, fertiliser, and cash advances. Yet farmers reduced their areas of double-cropped rice to grow cash crops which were more profitable. The quota price of rice and other grains was increased by 20 per cent in 1979, and the above-quota price raised by half as much again. Surplus grain could be sold on the free market. The high-yielding early maturing hybrids introduced in these years responded well to extra fertiliser inputs, and rice yields and production increased significantly. Inter-provincial trade had always been very difficult because of the limited transportation system, so government policy urged self-sufficiency in grain until the reforms of the eighties. In 1982 per capita consumption of rice averaged about 14 kg in the fifteen northern provinces, compared with 170 kg in the southern provinces. Conversely about 105 kg per head of wheat per year were consumed in the north compared with 18 kg in the south. But in the early 1980s the government proceeded to cut the size of the rice quota, and paid a higher price for it, giving the farmers the chance to produce extra rice above the quota. Rice output increased due to these policy changes and the greater use of hybrid varieties and HYVs. The State Foodgrain Bureau continued to control grain distribution. It channelled grains to urban residents and rural households who could not grow enough grain to support themselves, under a statutory rationing system. This also supplied at higher prices those who were not eligible for the rationing system. 1984 saw a good harvest, and that year China produced about 40 per cent of world rice output. But there was not enough storage for it, and the free-market price slumped. In 1985 the government reduced its acquisition at fixed prices. Commercial activities were also liberalised, and the government gave up its monopoly on grain procurement. Individual grain producers were allowed to sell a fixed amount of grain to the government at a unified 'contract price' determined as a weighted average of quota (30 per cent) and above quota (70 per cent) prices. Any rice in excess of the contract level could be sold to the government at a 'negotiated price' or directly to consumers or private traders. These were now allowed to obtain grain supplies from the free market.

Effectively the state stopped buying rice above the contract amount. The ration system continued as previously with grains distributed to urban consumers, farmers who could not produce enough grain for themselves, rural households which did not produce grain, and for disaster relief. The rations were set according to national standards for type of work, sex and age of recipients and sold in grain retail shops to coupon holders. Prices varied for each category. Urban residents received highly subsidised ration prices, at about two-thirds of contract procurement prices, but rural residents paid the same price as state contract procurement prices. Ration sales to urban consumers were much higher than sales to rural consumers. Sales made by the PFDS (Public Food Distribution System) were made to people excluded from the rationing system, to ration card holders wishing to buy extra or better quality supplies, to travellers and to grain-based industries. The prices of these grains were set at the negotiated price level, which was close to the free-market level. From 1984, following the removal of the state procurement monopoly, these sales of grain were used to stabilise free-market prices. In 1985 about 43 million tons of grains under the ration system were distributed to the cities at highly subsidised prices, compared with only 15 million in the countryside at procurement contract prices. Although the government's grain distribution policy was evolving towards a market stabilising operation, the rationing system continued to account for the larger share of public grain distribution. Since ration prices were considerably lower than procurement prices, these sales caused significant deficits, and imposed a considerable burden on state finances. The 58 million tons of grains distributed through the rationing system in 1985 were estimated to involve a total subsidy equivalent to a quarter of all state revenues. Later the procurement by the Foodgrain Bureau declined considerably. Also the government reduced the number of people who received rations, and consumers had to buy greater amounts at the higher negotiated level or at free-market prices. As part of the liberalisation process, private traders were permitted to market rice out of the southern provinces which usually had a surplus. Because government requirements had fallen, the cheap diesel and fertiliser which was given as part of the exchange for procured rice was also cut. New contracts for leasing land were reduced to three years or less, which meant farmers were not prepared to make any improvements on their land. In particular the irrigation network fell into disrepair, and this was made worse by the fact the government slashed its irrigation expenditure.

The Seventh Five Year Development Plan (1986–90) stressed the need to continue with liberalising reforms. Agriculture was to be the 'foundation of the national economy' and grain production was to be increased through increasing yields. But faced with a decline in rice production, incentives were again improved in 1986 by cutting the compulsory quota, and paying more for it, thereby giving the farmers a bigger surplus to sell. Farmers producing for the export market were supplied with better seed, low cost fertilisers, and prices well above the standard contract price. In 1987 the quota was cut again, and the price was raised yet again. The former practice was reintroduced of giving 60 kg of subsidised fertiliser and 30 kg of diesel for every tonne of quota grain, and advanced payments of 20 per cent of the contract price. The provinces were encouraged to increase their grain acreage. But by 1988 two years before the completion of the Seventh Plan, grain production had failed to meet the planned target for the fourth consecutive year. This was due to a combination of bad weather, and also low contract prices, high input costs and shortages of inputs. Free-market prices rose sharply in the south. Farmers were turning to other crops which gave them better returns. To encourage a shift from cash crops to rice, quota prices were raised, there was investment in irrigation and storage, and fertiliser distribution was re-centralised and increased. Interprovincial grain movement was banned again, to ensure that each province was self-sufficient in food. It was decreed that all rice quotas must be supplied in rice, and not commuted to cash as had been increasingly common. This prevented farmers from growing cash crops instead of rice. Restrictions were introduced to prevent farm land being converted to industrial use. These policies, plus good weather, led to good crops again. The government agreed to buy all the grain produced by farmers in specified areas, which meant the state had to provide increased storage facilities, even having to store rice in the open. In 1991 the government's procurement price for paddy remained the same for the third year in succession, which meant a drop in real terms because of rising prices in general. Financial restrictions limited fertiliser inputs and investment in land improvement, particularly water control. Storage problems continued with rice being stored in the open with high spoilage and losses.

The government recognised the problem of feeding its growing population, and the Eighth Plan (1991–5) intended to increase total food grain production from 425 million tonnes to 450 million tonnes and 500 million tonnes in 2000. Paddy output was to remain about 43 per cent of this total, or 194 million tonnes in 1995 and 215 million tonnes in 2000. The household contract system remained operative, with a maximum lease of 15 years, which was renewable. Production and procurement quotas continued, and subsidised inputs were related to the size of the quota, and included diesel, fertiliser, and plastic sheeting. A percentage of the crop was paid for in advance. Rice production would not be profitable at all if farmers had to buy their inputs at free-market prices, and sell to the state at the fixed quota prices. Only in years of good harvests was there much of a surplus to sell in the free market. The government planned to establish 'grain production bases' in which small scale irrigation and drainage systems were established and maintained, and the latest technology and information applied. Practices which conserved water and fertiliser were encouraged, as was the spread of hybrid varieties. More chemical fertiliser plants were planned (Barker *et al.* 1985: 60–1, 246–9; FAO 1991: 48–51 paras 3–7, 8, 10, 12: 56–7 para 31; Reynolds 1985: 278–80; Roche 1992: 34–6, 79–80, 150).

China was an important rice exporter in the early 1960s, but then exports declined. In the 1970s exports rose particularly to Vietnam because of the war, and also to Africa, Cuba and Sri Lanka. With the end of the war in Vietnam in 1975, the need to export rice there ceased. Exports to the rest of the world also declined in the following years, although there were considerable fluctuations. China did supply Cuba and Sri Lanka, but the rice did not come from the home crop and was obtained by the Chinese authorities from other countries. China continued to supply high quality rice to Hong Kong for much needed hard currency, and to Indonesia and Eastern Europe. During the 1980s China was again a net exporter of rice, and usually the third most important supplier to the world market. But exports in the early 1980s averaged only 670,000 tons, partly because Indonesia had cut its imports. Low world prices made exports unattractive at a time when China was trying to improve living standards at home. Exports to Eastern Europe also declined, but China continued to ship 190,000 tons of high quality rice annually to Hong Kong in 1980–4. Exports to Hong Kong continued in 1985–8, and there were also exports to Sri Lanka, Cuba, East Europe and West Africa, especially the Ivory Coast. There was some reduction in the exports to Hong Kong 1985–8, whilst exports to West Africa, especially the Ivory Coast, East Europe and Sri Lanka increased. Rice for Hong Kong and East Europe was usually of high quality, but exports to other countries were of 25 per cent and 35 per cent brokens.

Both China and Indonesia try to keep their crop yield secret to avoid weakening their position in the international market when buying or selling. In late 1988 they secretly bought heavily from certain international traders before the harvest was in, which left many other traders short of rice and pushed prices up in early 1989. The bad harvest of 1989 meant China was the world's largest rice importer that year. China's imports usually came from Thailand, and from Burma and North Korea. The Thai rice was usually low quality 25 per cent brokens. There were overland and coastal shipments from Vietnam. In 1989–90 there was a massive crop in China, but net rice exports were zero. There was another good crop in 1990–1 and the government raised rice procurement targets, and offered ready cash rather than IOU's for payment. Agricultural investment was increased.

China is also trying to improve rice quality, and their breeders have been trying to incorporate US characteristics into their rices. International trade in rice in China was administered by the Ministry of Foreign Economic Relations and Trade (MOFERT), the Ministry of Commerce and the Cereals, Oils and Foodstuffs Import and Export Corporation (CEROIL). CEROIL was the executive arm responsible for the import and export targets set every autumn by the State Planning Commission. These targets establish the quota of rice that CEROIL would buy from the Ministry of Commerce at the set price. Anhui, Jiangsu, Jiangxi, Hubei and Hunan were the main provinces providing rice for CEROIL's exports. CEROIL arranged rice export sales, but each shipment required a licence given by MOFERT. Imports were also CEROIL's responsibility, although this could be delegated to other bodies for implementation. During the 1980s there were attempts to liberalise foreign trade procedures. Provincial trading companies were allowed to negotiate rice purchases and sales in the world market, but they made losses and in 1989 were banned from trading in any kind of grain. CEROIL continues to control external trade, using international trading companies and brokers, and to a lesser extent government-to-government sales. In 1991 government-to-government sales to Eastern Europe almost halted, although up to that time Eastern Europe had been taking about a quarter of their sales. Since then China has come more into the international market, and the main category of rice is long-grained 35 per cent brokens. Export prices are usually below the prevailing world price. But with a more market based and commercial approach, and futures markets established in the main supplying provinces of Anhui, Hubei, Hunan and Jiangxi, prices may well rise. Export prices will also be affected by the prices of the subsidised inputs and without these subsidies, Chinese prices would be higher and probably uncompetitive with Thailand or India, although not the United States (FAO 1991: 49 para 1; Roche 1992: 35–6, 49–52, 108–9).

China's grain trade regime has become more complex in recent years, as attempts to move to a more market orientated economy have conflicted with the need for stable prices and grain self-sufficiency. Some new state corporations have been created, at times with opposing claims as to their power to import and export grain. In 1994 China National Cereals, Oils and Foodstuffs Import Export Corporation (CEROILS) continued to be part of the Ministry of Foreign Trade and Economic Co-operation system and was the main state trading enterprise engaged in the grain trade. Under the former centrally-planned economy, CEROIL alone handled the international trade in grains, oils and foodstuffs and essentially had only one customer, the former Ministry of Commerce. But recently provincial branches of CEROILS and other groups have been allowed to trade in certain commodities. CEROILS still controls trade in wheat, corn, barley, and soybeans and has regained control over the vegetable oil trade. CEROILS supplies the new Ministry of Internal Trade, which replaced the Ministry of Commerce, and makes imports for provincial, municipal and other buyers. Normally CEROILS takes 100 per cent cash in advance from these customers, but it also can buy on its own account if requested by the government. The old Ministry of Commerce became the Ministry of Internal Trade (MIT) and handles domestic distribution, and the processing and marketing of grains, oils and other products. MIT controls thousands of flour and feed mills, breweries, and other food processing units, and has sought to get some control over imports of grain and oilseeds. Under the old centrally-planned system, MIT had virtually no control over the quality or origin of supplies delivered by CEROILS. But from 1992

MIT has moved to increase its role in the grain trade. The government established the State Administration of Grain Reserve (SAGR) to buy and sell grain to stabilise the domestic food and feed grain markets. This was during the early attempts to seek a more open grain trading system and increase competition in flour milling. SAGR is linked to MIT, and whilst it buys domestic grain and controls about 40 per cent of China's wheat imports, it does not itself make grain imports. SAGR has recently made a commercial agreement with CEROILS to establish the China National Liang Feng Grain Import and Export Corporation, whose aim will be to operate in the international grain market to fulfil SAGR's aim of stabilising domestic grain markets. Another firm recently approved by central government in the MIT system is the Zhong Gu National Grain Enterprise Holding Corporation. This is to be operational in flour milling, food processing, retail sales and domestic grain marketing, and will be able to purchase some of China's wheat imports. Despite these changes CEROILS retains tight control over China's grain trade. Although those seeking grain may now specify the origin and quality of the grain they need, they cannot themselves purchase and make trading profits. Price stability is China's main concern, and the desire to maintain their reputation as an international buyer and seller indicates that CEROILS will continue the main trading role. CEROILS also has to ensure grain self-sufficiency. The character of China's state trading enterprises, especially CEROILS has been a subject of discussion in the negotiations concerning China's entry to GATT and the World Trade Organisation. CEROILS monopoly power to supervise China's grain trade creates problems for US agricultural exports. But China will be unwilling to undo this system which enables them to manipulate trade, ensure price stability, and control foreign exchange, unless there are good reasons for doing so. However the costs to consumers of maintaining this system to ensure grain sufficiency and stable prices may ultimately lead to change (*Grain* 1994 November: 12–14).

The attempts to introduce market reforms in agriculture ran into difficulties in the mid-1990s, as it was thought these reforms had led to inflation. It was believed that with less grain held by the State Administration for Grain Reserves and the Ministry of Internal Trade, the state was unable to hold down prices, and corn and rice exports were therefore banned in 1994. It is still not clear how grain supply and demand balances in China, and if the government will recognise that there is a gap between production and demand (*Grain* 1996 March: 10–11). But since 1994 there has been a transformation in China's role in the international rice market. In 1993 China was the world's fourth largest exporter of rice, but in 1994 there was extensive flooding in the south with devastating consequences (*LRBAC* 31 July 1994: 1; 30 September: 1, 2). Imports increased sharply, and exports increased as well, leaving China with net exports of over 800,000 tons, although she remained the fourth largest exporter. But in 1995 China was the world's second largest importer of rice, and exports virtually ceased. Imports seem likely to continue, and many food analysts believe that they are likely to continue because of population growth of more than 10 million a year, land lost to industry, and general per capita increase in grain consumption. However, China has been a rice importer in every year since 1960 except three. China's imports were due to a shortfall of domestic production of rice of only 1 per cent, and represented less than 10 per cent of world trade, although this did have an impact on international prices. Clearly the size of China's rice market means that a shortage in any year will have a major impact on world markets (*Grain* 1997 April: 14–15).

Previously China was the leading supplier of low quality rice to the world market, but in recent years it has become a major importer of high quality rice, including fragrant and 5 per cent brokens. In 1994 when Chinese imports were over 700,000 tons, 400,000 tons were best quality Thai rice. Even in 1995 there were 700,000 tons of high quality rice from Thailand and Vietnam, when total imports soared to 2.0 million tons. 1995–6 saw very good crops in China, so low quality imports ceased, but premium quality imports from Thailand and Vietnam continued at a high level. Between January and April 1996 China imported 24 per cent of Thailand's exports of high quality rice, and a third of all Thai exports of fragrant rice. These imports have been a major factor behind high international prices for premium quality rice. Hence the disparity of $100 per ton in July 1996 between 100B (the key grade for high quality Thai rice) and 35 per cent broken rice in Thailand. In consequence many believe that China's demand for high quality rice will make the market for such qualities very tight in the future. Demand for these qualities is also growing in Europe, North America, and other prosperous Asian nations, at a time when Thailand, Australia, India and Pakistan seem to have reached a limit on land available for rice cultivation. Some increased supply of these qualities will result from improved milling and processing, and here Vietnam has obvious potential. More fragrant rice however could be planted instead of inferior qualities (*Grain* 1996 July: 9–11).

In June 1996 China shocked the international rice trade by winning the tender for the South Korean minimum access obligation under the Uruguay Round, and followed it by capturing more than 50 per cent of the 5,000 tons of an initial minimum access contract for Japan. United States exporters who had dominated Japan's 1995 imports with a 48 per cent market share got only 1,200 tons in sales, Australia had 800 tons and Thailand 119 tons. This tender accounted for less than 1 per cent of Japan's minimum access quota, but it was clearly a threat to the US position, and to many people signalled the likelihood that China would resume large scale exports (*Grain* 1996 August: 10).

North Korea

In 1995 and 1996 North Korea saw heavy flooding disrupting domestic production and cutting it by about a half. There were gifts of grain from South Korea in 1995, and in that year and in 1996 there were also donations from Japan. Domestic production averaged 4.7 million metric tons 1985–6 to 1996–7 but in the last two years harvests were poor, and stocks dwindled to nothing. In 1997 the food ration for adults was cut from 500 g to 300 g per day. The UN World Food Programme appealed for $41.6 million to feed North Korea's desperate population, and the US government offered about $10 million. South Korea offered a further $6 million. The $41.6 million was to buy 100,000 tons of maize, rice and maize-soy mix. North Korea's twenty-four million people need 4.3 million metric tons of grain a year, plus another million for livestock feed and industrial uses. In the past North Korea has relied on domestic production, plus imports from China. But China has not so far openly offered relief supplies, although they have been a traditional supplier. Nor has Japan offered any food help either, although they have a large rice surplus available if the government decides to assist. But there will still be a shortfall, and North Korea will have to turn to the international commercial market. Barter negotiations have begun with US Cargill, but as North Korea has little cash, and a poor credit rating, prospects do not look good (*Grain* 1997 March: 10–11). The *Daily Telegraph* reported in April 1997 that there was 'slow starvation on a massive scale' and the US State Department estimated that 6–8 million people were at risk of starving to death (*Daily Telegraph* 9 April 1997: 16; 6 May 1997: 14).

South Korea

South Korea only became a major rice producer and consumer in the 1930s, under the influence of the Japanese colonial administration. After the Korean war (1950–3) South Korea aimed for self-sufficiency by giving substantial subsidies to farmers and restricting imports. Self-sufficiency was attained in 1984. Since then it has faced a situation rather like that of Japan. Government support has led to too much rice being produced at a time when rice consumption was falling because of rising incomes. Like Japan, Korea produces round-grain japonica rices, and could become an exporter of these grades should government policy and world demand permit (Barker, *et al.* 1985: 56; Roche 1992: 130; Lee 1990: 21). In 1993 Korea committed itself under the Uruguay Round GATT agreement to allow imports of rice in 1995 of about 50,000 tonnes rising to about 200,000 tonnes by the year 2005, a rise from 1 per cent to 4 per cent of annual consumption (*LRBAC* 31 December 1993: 3). So in 1995 South Korea's Office of Supply (OSROK) bought all its 51,000 ton GATT minimum access commitments from India, the cheapest bidder, because it wanted to pay as little as possible to fulfil its obligations. The rice was long-grain rice only suitable for industrial use. In 1996 OSROK again ignored US rice for its GATT minimum access purchase requirements, and took all its required imports of 64,000 tons from China. The final $442 per ton bidding price was more than $100 per ton below the Australian and American bid prices. However the quality was suitable for table use. Forecasts for Korea's 1996–7 harvest suggest it will be the sixth in a row which fails to provide enough for domestic use, and that South Korea will need to import up to 500,000 tons, well above its minimum access commitment of 77,000 tons (*Grain* 1996 July: 11).

Taiwan

Taiwan exported an average of 120,000 tons a year in the early 1990s, and has in recent years tried to introduce market forces to its agricultural economy. Up to 1995 the government undertook to purchase two crops of rice a year from farmers, but under proposed changes this was to continue just until 1997 when purchase of only one crop per year would be guaranteed. There was also guaranteed purchase of corn, sorghum and soybeans produced on land converted from rice production, but this too was to be limited to one crop a year in 1997, and cease from 1999. The aim was to reduce rice production by about a third, and take Taiwan from the export market, generating a low level of import demand (*Grain* 1994 December: 12).

Japan

Rice is grown throughout Japan, and an early hybrid was produced in 1898, with twenty hybrids being grown by Japanese farmers in 1913. The Japanese tried to control the rice market as early as 1921, seeking to increase production whilst simultaneously keeping prices down. In response to the depression of the 1930s they imposed duties to keep out cheap foreign rice, and began a policy of government purchasing to raise prices and increase domestic production. Soon they had a rice surplus and chronic warehousing problems! In 1939 the government took control of all rice markets, banned speculation in rice, and restricted trading to licensed brokers and dealers. In the years after the Second World War the Japanese economy was greatly disrupted, and they became a major rice importer. But as economic recovery took place and incomes rose, per capita rice consumption began to fall as meat, fruit, vegetable and bread was added to the diet. Low rice prices were no longer an essential aim, so the government began a programme of income support for farmers to ensure that self-sufficiency could be attained. The necessary subsidies could now be afforded because of increased government revenues. Between 1958 and 1969 prices paid to rice farmers doubled, and production increased at a time when domestic consumption was decreasing, making imports unnecessary. Production has been controlled and subsidised by the Japanese Ministry of Agriculture, Forestry and Fisheries (MAFF) since 1971. As consumption fell, production was cut systematically, although a surplus was still produced which is not normally exported. In 1978 farmers were receiving $1,100 million per tonne, whilst export prices for Japanese rice were less than $300 per tonne! Exports were only made as subsidised concessions and depended on government policy and the warehouse stock situation. Although rice consumption was declining, there was a trend towards higher qualities of rice, and rice was still required for animal feed and confectionery. Up to 1987 MAFF encouraged farmers to move from rice to other crops. Each local co-operative was given a quota, which was shared amongst the farmers. Each farmer then gave his crop to the co-operative, and 85 per cent of the total crop went to the MAFF Food Agency. The rest was retained on the farms. The Food Agency has a monopoly on rice exports, although these are rare. In 1990 the rice organisation in Tokyo and Osaka began dealing in over forty varieties of rice. Control of the Tokyo and Osaka exchanges rests with the professional body Kakaku Kikoo, which is under the general supervision of MAFF. There are two official prices. There is the official production purchase price decided annually by MAFF, the government, and the Rice Price Advisory Council (RPAC). The government has tended to raise the price proposed because of the influence of the farming lobby. The other official price is determined indirectly through other prices, and since 1986 has been below purchase price. Both prices have been falling for some years. This system results in exceptionally high rice prices, and is costly both to the state and to the consumer. The US, Australia and Thailand have long pressed for access to the Japanese market. To ensure rice self-sufficiency for strategic reasons, Japan has in the past prohibited rice imports even though the price of rice in Japan was up to ten times above the world price. In 1986 the US Rice Millers' Association objected to the rice embargo, and it became a major issue at GATT negotiations. The United States Department of Agriculture took the view that the Japanese exaggerated the idea that there would be a flood of rice into Japan if the embargo was lifted (Barker, *et al*. 1985: 54; Roche 1992 : 38, 70–5, 129–30; Wickizer and Bennett 1941: 165–87).

In 1993 Japan experienced the worst conditions for growing rice for almost 50 years, with cold wet weather lasting well into the growing season (*LRBAC* 1 October 1993: 1). It was this year that Japan committed themselves under the Uruguay Round GATT agreement to become a regular importer of rice, buying about 400,000 tonnes in 1995 rising too about double that by the end of the century (*LRBAC* 31 December 1993: 2; Tsujii 1995: 131–2). The harvest in 1993 did turn out badly, and the outcome was that Japan had to import 2.5 million tons of rice in 1994 to maintain food supplies and conserve stocks. This was the first time Japan had imported rice for many years, because high prices supported by the government meant that farmers produced enough for domestic needs. But much of the rice was of varieties unsuited to the Japanese market, and there was a short term shift to noodles and wheat based alternatives. So when the 1994 harvest turned out to be a record post-war harvest, much of the imported rice was allowed to remain in store. The government stocks stood at 800,000 tons, made up of 480,000 tons from China, and 250,000 tons from Thailand. How these surplus stocks could be disposed of was a problem, and much was to be sold for animal feed, with some to the food processing industry. There were food aid shipments to Laos and Nepal, and North Korea was also interested in obtaining large amounts as aid, although diplomatic relations did not exist between the two countries. Japan also needed to reduce its rice stocks because of its Uruguay Round GATT agreement of 1993 to establish an import quota for 379,000 tons of rice in 1995, increasing to 758,000 tons in the year 2000 (*Grain* 1995 June 20; *LRBAC* 31 December 1993: 2). During Japan's emergency rice import campaign of 1993–4 China was the leading supplier with 1.1 million tons, more than 40 per cent of the total. Thailand supplied 785,000 tons or 30 per cent of Japanese imports, and Australia 192,000 tons or 7.3 per cent. US exports of 548,000 tons or 21 per cent were limited by a shortage of suitable medium grain rice in California. But in 1995 there was no crisis and both the US and Australia were able to supply the market having developed varieties suitable for Japan, and the processing facilities necessary for the high quality product required. In both years the government of Japan's Food Agency (FA) made the necessary purchases, and transport and storage was handled by licensed Japanese importers. The rice had to be subjected to a strict three part examination to meet Japanese safety and hygiene norms (*Grain* 1995 November: 8). In 1995 the FA imported 408,794 tons to fulfil Japan's obligations under the Uruguay Round minimum access requirements. This sum included both brown and fully-milled rice. The US share was 194,000 tons (47.4 per cent). Australia was awarded 87,000 tons (21.3 per cent), Thailand 95,000 tons (23.3 per cent), China 32,000 tons (7.9 per cent). Small quantities were also purchased from Pakistan and Uruguay. These imports fulfilled the minimum access requirements for rice imports by Japan to the end of Japan's fiscal year ending 31 March 1996 (*Grain* 1996 January: 7–8).

Australia

Rice cultivation began in Queensland in 1869, the rice being dry upland rice. Hundreds of acres were cultivated between 1880–90, but then sugar took its place. Rice was also tried at Darwin in the Northern territory. There were attempts to grow dry rice in New South Wales from 1891 without much success, and wet rice also proved unsuccessful because unsuitable varieties were used. The first successful rice experiments were undertaken in the Murray River district about 1914 by a Japanese immigrant, I. Takasuka, who developed a variety which bears his name. Rice cultivation began in earnest in the Murrumbidgee Irrigation Area of New South Wales in 1924–5, and has continued to be a major crop. In 1954 large scale cultivation began in the Riverina district there using mechanised equipment, and the highest yields in the world have been obtained. This is due to good control of water, excellent soil conditions, plenty of sunshine, and a good rotation of rice with other crops, including wheat, oats and leguminous pasture plants for raising fat lambs. Rice also grew in the Northern Territory (Grist 1986: 11, 183, 289, 492, 508; Hungerford 1950: 498–501; Black 1996; Lewis 1994). In the late 1950s a major rice growing scheme was instigated by Territory Rice Ltd at Humpty Doo about 40 miles east of Darwin in the Northern Territory, but it collapsed in 1962 after predation by geese and water buffaloes. There were several US investors in the project. The motivation behind this scheme appears to have been to secure rice supplies for Asia in case the major rice growing districts of Asia fell into communist hands. This did not happen, but Australia has emerged as a supplier of significance, with supplies coming from the irrigated Riverina district of New South Wales (*The Australian* 5 January 1984: 1, 4). The rice industry in Australia centres on Leeton, New South Wales (see Appendix).

United States of America

Before the Second World War the United States had been a relatively minor exporter to world markets. But after the war the US government encouraged increased rice production because of the world food emergency. Rice production doubled in the US between 1946 and 1954. That year the Korean War ended and there was a surplus for the first time. So the following year acreage controls and marketing allotments were introduced. But world and domestic demand for rice continued to expand. Public Law 480 (1954) established that federal farm surpluses stored with the Commodity Credit Corporation and similar organisations could be issued for sale to developing countries in their own currencies, and to relief agencies and for school lunches. The aim was to reduce surplus stocks without depressing world prices. Between 1954–7 the government tried to reduce acreage and cut home production to eliminate overproduction by halving rice allotments. In 1958 a bill was introduced in Congress providing for payment-in-kind by which rice and other American commodities could be exchanged for foreign goods which were in short supply in America. These included tropical items such as rubber, cocoa, spices, hemp and some dyes. The payment-in-kind principle expanded the world market for American rice considerably. In 1959 there was an amendment called 'Food For Peace' which meant the US could lend countries money on very favourable terms to buy food. Rice was a key food in the 'Food for Peace' programme, so the effect was to increase world demand for US rice. That year a PL480 agreement was made to supply India with 1 million tons over four years, which marked a further expansion of American rice into world markets. This was fortunate for American rice producers, because Cuba had fallen under Castro and soon ceased to be a market, taking rice instead from China. In 1961 large export contracts were signed with Vietnam and Indonesia, and these markets grew rapidly. South Korea also became a purchaser of American rice, and sales to the United Kingdom, the European Economic Community, South Africa and Japan also increased. Soon the United States was vying with Thailand as the world's largest rice exporter, with sales of over 1.5 million metric tons in 1966. American rice shipments to South Vietnam under PL480 rose from 0.6 million cwt (30,000 tons) in 1965 to 4.3 million cwt (215,000 tons) in 1966, but ended when the South fell to North Vietnam. However, in the 1970s, revolutions and wars in Iran, Iraq and Pakistan gave new opportunities for American rice, and North African and Middle Eastern countries increased their purchases of American rice as they profited from an oil embargo. During this decade 90 per cent of Saudi Arabia's rice was supplied by American Rice, Inc. Much of this success was due to the activities of the Foreign Agricultural Service (FAS) of the United States Department of Agriculture (USDA), which has market development offices in major European market centres and South Africa, and also collects and collates information from US embassies and consulates abroad. FAS also co-works closely with the rice industry's trade association, the Rice Council for Market Development (Dethloff 1988: 162, 171–83) African and Middle East markets continued to be the leading customers for US rice, in the 1980s. Iraq was a major destination, and the loss of this market after 1990 meant a search for other buyers. USDA found new markets in countries of the old Soviet Union, Czechoslovakia, Hungary, Poland, Romania and what was Yugoslavia. PL480 and EEP schemes (see below) were used to make these deals. There were also sales to Turkey, and relief supplies given to Jordan to help refugees from Iran and Iraq. The Russians too bought rice. USDA has programmes for commercial export assistance, and they include two export credit guarantee programmes, four export subsidy programmes, and two programmes providing promotion and trade servicing to maintain and expand markets. These schemes help sales made by US private companies selling at market rates. But USDA also has two export guarantee programmes which assist US exporters sell to developing countries and countries who can not raise hard currencies. The aim is to wean countries from taking the special cheap rates under PL480 on to normal commercial trading, or to help them over economic crises. Current interest rates are charged. The Export Guarantee Programme (GSM-102) ensures USA exporters are paid in the event of default by a purchaser. The exporting country pays a fee to USDA's Commodity Credit Corporation (CCC) before shipment. The foreign purchaser buys the rice (or other commodity), and obtains a letter of credit from a bank. Repayment is over three years at current commercial rates. The CCC will cover these deals if they think US trade will be benefited, and if they think the risk is sufficiently low. GSM-103, the Intermediate Export Credit Guarantee Programme, is like GSM-102, but guarantees credits for up to ten years, rather than the three years of GSM-102. Although these schemes are available, they are used more for other commodities than rice. There is also the Export Enhancement Programme, by which US exporters can sell at world prices, and be paid an additional subsidy by USDA.

The Food, Agriculture, Conservation and Trade Act of 1990, administered by USDA FAS provided for $200 million annually in 1991–4 to help US producers promote rice and other agricultural products, as authorised by the Rice Council for Market Development. Houston, Lake Charles and New Orleans on the Gulf of Mexico are the leading US rice ports, handling rice from Texas, Louisiana and Arkansas, and Oakland on the West Coast handles California rice (Roche 1992: 41, 111–16, 197–8, 204).

Until the 1970s US rice production was mainly medium and short-grain varieties. Long-grain rice was less than 45 per cent of total production. Rice growing had been started in California for Chinese and Japanese immigrants, and as late as 1980 California was producing 25 per cent of US rice production. Most was of the medium- and short-grain varieties which are suitable for Japan and Korea. Japan needed to import rice in the years after the Second World War, and Californian rice found a ready market there up to the late 1960s. When exports to Japan declined the rice went instead to Korea which found themselves with rice shortages. South Korean consumers like the same medium-grain rice as the Japanese, so the Californian rice was ideal. But the short-grain rice which made up about 15 per cent of US output in the late 1960s was only taken by the Japanese, and loss of the Japanese market resulted in a collapse of production. The early 1970s saw continued steady growth of production of long-grain rice in the US and their international aid programmes helped provide markets for this rice in Asia. The PL480 programme gave United States government funds to foreign governments to buy United States agricultural products. These products would then be sold by the foreign government to their people, and the revenue used to fund development projects. There were large PL480 projects to India, Indonesia, and the Philippines. Indonesia received the most, enabling her to purchase imports of US rice totalling approximately 400,000 tons in 1976 and 1977. PL480 rice and commercial exports of rice to South Korea meant that United States exports of rice to Asia regularly exceeded 400,000–500,000 tons in the late 1970s, and in 1980 and 1981 were about 1 million tons a year. Exports of US medium grain to South Korea in 1980 were 800,000 tons or 30 per cent of total US rice exports to the world at large. PL480 allocations for long-grain rice were cut back drastically in the 1980s, and as South Korea became self-sufficient US rice exports to Asia virtually ended. In 1981 they had been 1.0 million tons but by 1984 they were down to 90,000 tons. These levels continued for more than ten years, apart from heavy PL480 shipments to the Philippines in 1988. In 1981 United States exports were about half of total imports of rice in east and south-east Asia. But the US share then fell, as Asian rice importers purchased more and more from producers within the region. Thailand maintained a steady share of the market, but China, Vietnam and recently India began to supply their neighbours. But there has been revival of US medium-grain exports to Asia, and if market liberalisation continues in Japan, and production declines in South Korea prospects for US exports are likely to improve. The US exported over 800,000 tons of medium-grain rice to South Korea in 1980 and 1981 but in 1982 exports fell to just over 200,000 tons. By 1983 total US exports to all markets including South Korea were only 291,000 tons, less than 30 per cent of the levels of 1970. Medium-grain exports were down to 10 per cent of the US total export figure. As export demand fell, so did domestic prices for medium-grain, down from $27.70 per cwt in 1980 to $15.90 in 1982, a fall of 40 per cent. Long-grain prices also fell, but not so severely, and whereas medium-grain had sold at prices above long-grain, now the situation was reversed. So American farmers cut medium grain production by more than 45 per cent in 1982–3 both in the south and in California, and although planting soon recovered in California it did not do so in the south. Instead long-grain production developed rapidly. Until the 1970s, long-grain had provided less than 50 per cent of the crop, but by 1985 it was 70 per cent. However, medium rice production began to recover in the mid-1980s for industrial use and food processing. From 1986 prices for medium-grain began to rise again. Although medium-grain exports to Asia had virtually ended, new markets were beginning to emerge in sub-Saharan Africa and South America under the PL480 programme. By 1986 Middle Eastern markets were the leading importers of US medium-grain, with Turkey the largest single importer, with purchases of 179,000 tons in 1989 rising to 300,000 tons in 1995. Jordan too became a steady buyer, with an average of 45,000 tons in 1986–92.

Then planting of medium-grain rice began to revive in the southern states, and by 1990 had overtaken California, and total US production of medium-grain that year was 2.2 million tons. In 1983 Asia again became an export market with Japan's emergency import campaign leading to US exports of 30,000 tons in 1993 and 481,000 in 1984. FOB prices for No. 1 Calrose medium-grain rose to $600 per ton by late 1993, more than twice the price in the first six months of the year. The cultivated area of medium-grain rice has continued to expand. Whilst Japan was driving prices up to record levels, US exports to the Middle East in 1993 and 1994 continued strong. As Japan began its GATT minimum access purchases in late 1995 and 1996 medium grain rice prices remained high. Turkey however also had record imports from the US in 1995, reaching 300,000 tons, and there was strong demand from Jordan. There is considerable potential for US exports to east and south-east Asia, as imports to the region have increased from about 2 million tons to 5 million tons per year. Indonesia, China and the Philippines have reached cultivable limits, and yields are stagnating at a time when demand is growing. Uruguay Round commitments will require Japan to import three quarters of a million tons of rice by 2000. Similarly Uruguay Round commitments and declining production in South Korea indicate heavy imports in the future. Opportunities for US long-grain exports to Asia are limited, and the market is for premium quality medium-grain. Singapore, Hong Kong and Indonesia are also markets for the kinds of high quality rice imported by Japan. Medium-grain production in the US is expected to rise, fulfilling increased demand in Asia, the Middle East, Canada and Europe. There will also be domestic demand from industrial users and food processors (*Grain* 1996 November: 9–15).

Indonesia has made huge rice imports from US, Japan has made their minimum access purchases, and China has changed from being an exporter of low quality rice to an importer of high quality rice, so much attention has been concentrated on the Asian market for US rice. But whilst the development of markets in the Middle East and Europe have been important for US exporters, it is the growing demand in Latin America which has had the biggest effect on the US rice industry. Demand for rice imports in Latin America has risen from 600,000 tons in 1988 to an anticipated 3 million tons in 1997, a fivefold increase. Mexico, Peru, Haiti, Brazil and Central America have all become important regular purchasers. This is due partly to the regions strong economic growth, and partly to the fact that industrialisation and agricultural change in Mexico, Peru and Brazil has resulted in falls in domestic rice production. The US has freight advantages in Latin America, and also tariff advantages in some countries. But the main reason Latin America has become the key export market for the United States is the fact that only the US and Argentina export paddy, or rice in the husk. Other countries export milled rice. Argentina's paddy goes mainly to Brazil, but US paddy is exported to all Latin American countries. US paddy or rough rice exports are now 16 per cent of total US rice exports, about 7 per cent of total US rice production.

American millers do not like paddy exports, for they bring foreign millers who buy in the US market in direct competition, pushing up US paddy prices. Also, there is surplus milling capacity in the United States, and the millers have to cut production towards the end of the season. But Latin American millers can keep their mills in operation after they have milled their domestic crop by importing US paddy. So US millers argue that exporting paddy enables foreign millers to steal milling capacity from them. If Mexican or Central American mills were to close, those countries would have to buy US milled rice. However, the fact that Latin American millers rely on US paddy means that they have to mill and package it at a price which is competitive with US millers. Virtually all rice milled in Mexico now comes from the US and in Central America the figure is 25–30 per cent of the total. But millers in Mexico, Central America and South America cannot mill US paddy and sell it at a price which is competitive with imported milled Asian rice. So their governments have moved to prevent the import of Asian rice, thereby protecting their milling industry. Even US milled rice would not be competitive in these markets with rice from Thailand and Vietnam. In 1996 US exports of milled rice (as apart from paddy) was over 200,000 tons. The ban on Asian rice in many markets of the region allows the US to sell in these markets without serious competition. This is remarkable because Mexican mills now mill more US paddy than Mexican paddy. Under NAFTA, tariffs on rice imports from the US will be eliminated in the next six years, and Mexican millers will then have to compete directly with US millers, whilst bearing the extra costs of importing paddy. Many Mexican millers may go bust, and then the case for keeping Asian rice out of the Mexican market will no longer exist. US millers need to prepare for the possibility of direct competition with Asian suppliers by establishing consumer loyalty to their rice whilst Asian rice is still excluded (*Grain* 1997 March: 8–10). Whilst continuing to be a major rice exporter, the US has also become an importer of Asian rice since the late 1980s. Most of it has been fragrant rice, such as jasmine rice from Thailand and basmati from India and Pakistan, and these are qualities which the United States does not produce. They are long-grain rices, but their fragrant characteristics distinguish them from normal long-grain rices of the kind which the US does produce. Consumption of these grades has been growing rapidly, and most of the increase in US rice consumption is of these and medium-grain varieties (*Grain* 1994 November: 8).

In August 1994 the US rice industry decided to bring together all its major sections to create a single national organisation to represent the industry and co-ordinate resources. The USA Rice Federation was composed of the USA Rice Council, the Rice Millers' Association and the US Rice Producers' Group. Programmes and administrative duties for all three organisations are now conducted by USA Rice Federation staff, although each member organisation retains its own board of directors. Each member organisation has six representatives on the USA Rice Federation's Board of Directors (see Appendix).

Brazil

Brazil is the largest consumer of rice in the Americas and rice is their traditional basic food. Most rice is sold in supermarkets in 1 kg packages, ready for cooking. Brands linked to particular mills are important, although national brands are now emerging, such as Supra-arroz, which takes nearly 5 per cent of all sales. There is a strong consumer preference for the cooking qualities of Brazilian rice. Imports are usually made by millers who process and package the rice. Because of the preference for Brazilian qualities, rice imported from Uruguay, Argentina or the United States is not labelled as imported, and is packaged under the same brand names as home produced rice. Although the millers prefer to use domestically produced rice, they do need to import from time to time, and therefore maintain good relationships with exporters in Argentina and Uruguay, many of whom are themselves Brazilians. There are two rice growing regions in Brazil. Dry rice is grown in central Brazil on land recently deforested and about half the crop is produced in this way.

Although deforested land loses its fertility after a couple of seasons, the pace of deforestation means that there is always new land to plant to replace the exhausted areas. So production from this area remains stable. But in the other rice growing region, the Rio Grande Do Sul, production is falling. Here rice is produced by modern mechanised methods from levelled irrigated paddy fields. But farmers are being driven out of business by their debts and lack of government support, many moving to northern Uruguay and north-east Argentina where costs are lower. In consequence domestic production has been falling since the 1980s. But Brazil has been able to obtain most imports from their neighbours in the MercoSur free trade area, particularly Argentina and Uruguay where production has been rising assisted by farmers newly arrived from Brazil itself. Brazil has been the leading rice importer in America since the 1970s, and their annual imports now are close to 1 million tons. This makes Brazil one of the world's major rice importers (see Chapter 7) (*Grain* 1996 October: 10–16).

Uruguay

Uruguay has been a substantial rice exporter since the 1970s. Rice production in Uruguay used to be mainly from the Laguna de Merin region in the south-east, close to the Atlantic coast. But the area under rice has increased by over 75 per cent in the last decade due to expansion in northern Uruguay, much of it undertaken by farmers who have emigrated from Brazil. Production has doubled, and exports have risen from 200,000 to 500,000 tons, helped by Brazilian demand. The quality of rice is very good, and the equivalent of US rice. Production seems likely to expand, as they have suitable land which is unused. However, expansion will depend upon provision of irrigation facilities, which imposes high development costs. All Uruguay's rice is produced by advanced modern methods, with extensive irrigation structures. There are only two large scale exporting companies, and they maintain close links with the farmers. They buy the paddy direct from the farms, and mill, process and market it. These processors even own direct watering systems, so farmers are dependent upon them not only for their sales, but their water as well! The variety Blue Belle has been the most favoured Uruguayan rice, but although the processors have paid a premium for it, the premium has been insufficient to offset the fact it is low yielding. As a result farmers have been turning to higher yielding Brazilian varieties, especially Paso-144. However, these new rices are not welcomed in some export markets. This is something of a problem, as Uruguay is actively seeking new export markets, such as Turkey, Iran, Cyprus and Senegal. However Brazil itself will remain their main market for the foreseeable future, although they also export to Chile and Peru (*Grain* 1996 October 10–16).

Argentina

In the past Argentina was only a minor rice exporter, but in the 1990s exports have risen from 75,000 tons to nearly 400,000 tons, mainly destined for Brazil. Now rice is also going to Peru and Chile. The area under rice has more than doubled in the last decade and if market conditions remain favourable, is likely to double again in the next ten years. There is plenty of available land in north-east Argentina, and ample water, but investment will be required in processing facilities and storage. There is heavy Brazilian investment in the Argentine province of Entre Rios, which is linked by rail to Sao Paulo, giving easy access to the Brazilian market. Most of the new growing areas are planted with Brazilian rices such as Paso-144, suitable for Brazilian consumers. Despite these new developments, about a third of the domestic crop is consumed in Argentina itself, the rest being exported. There are many small scale mills servicing the domestic market, and the largest rice miller in the country – La Arrocera Argentina – only handles about 10 per cent of the crop. Because there are many small millers Argentinean rice is not of a consistent standard when it is marketed, although there is no problem with the quality when the local paddy is milled by a reputable company. Unfortunately no such firm is active in the export market and exports to Chile and Peru have a reputation for variable quality. So Argentina retains considerable potential as a rice exporter, but it will probably require foreign investment in milling and processing capacity to develop it (*Grain* 1996 October: 10–16).

Peru

Rice and potatoes are the staple diet in Peru, and Peru has become one of Latin America's principal rice importers, taking 250,000–350,000 metric tons a year since 1989. Most Latin American rice importers tend to take their rice from one source, but Peru imports from the US, Uruguay, Vietnam and Thailand. After ten years of political disorder, Peru has liberalised its economic system, and the rice industry has benefited from the closure of the state trading agency (ECASA) and the privatisation of the rice trade. In the past rice has been sold by shopkeepers filling bags from 50 kg sacks and weighing them for individual customers. But Lima and Arequippa now have supermarkets, and prepacked 1 kg bags take about a tenth of sales and are increasingly the norm. Rice imports to Lima arrive by shipload, which are then 'broken' into small lots of 50 kg sacks and sold to shopkeepers. This means that at any one time the shops tend to have rice from the same ship, and this determines the price level in the market. However, the difference in quality of rice from different countries is well recognised, and a hierarchy established. Pakistan and Indian rice is regarded as of poor quality (55 Soles per 50 kg). Sales of these grades are small despite the low price. Vietnamese 5 per cent, Thai 10 per cent and home grown rice are regarded of medium quality (66 Soles per 50 kg), and rice from US, Argentina and Uruguay are high quality (90–5 Soles per 50 kg). Although Thai rice is much cheaper, US rice sells well, but sales are affected by variations in the price of competing Argentinean and Uruguayan rice. Three major companies import rice of medium and high quality for brand name packaged rice. The medium quality packaged rice uses local rice mixed with rice from Argentina or Uruguay, but high quality packaged rice uses only rice from the US, Argentina and Uruguay. The firms import rice, polish it and pack it. They prefer to import brown rice or even paddy, as this way they can control the milling quality more effectively. Such rice retails at 40–50 per cent more than similar grades sold loose from 50 kg sacks. It seems possible the price of packaged rice can be brought down by bulk buying entire shiploads, and the sale of by-products like bran for feed and bran oil. Because packaged rice is more expensive to the consumer than loose rice from the sack, sales of packaged rice have not grown very rapidly. The packaged rice sector is likely to give US exporters the best chance of increased sales in the Peruvian market. The packers require a steady supply of rice all the year round, and the US can supply this, whereas Argentina and Uruguay can only supply at certain seasons. Also the US suppliers can guarantee quality levels as they have grading criteria and an excellent inspection service, unlike their main competitors. Media promotion in Peru is good, and this could be utilised by those promoting US rice. There are also many US exporters for Peruvian importers to chose from, but only two in Uruguay, and one in Argentina. Brazil has recently increased imports from Argentina and Uruguay, so planters in those countries have begun to plant Brazilian rices instead of the traditional Blue Belle variety. But the Brazilian rices are regarded as inferior in Peru (*Grain* 1996 December: 13–15).

7

TRADE IN THE 1990s

An examination of trade in the 1990s is shown in Tables 7.1 and 7.2. Possibly the most important finding is that world trade in rice nearly doubled between 1990 and 1996, from 11 million metric tons in 1990 to nearly 21 million tons in 1995.

If the characteristic of the world export trade is its concentration in a few exporters, the opposite is true of the import trade. Table 7.3 lists the imports of twenty-four of the leading importing countries, but Table 7.4 reveals that even they take only about 60–70 per cent of the total leaving many other importers unlisted. Whereas nine exporters supply 90 per cent of world exports, some twenty-four importers take 60–70 per cent of imports. In other words there is a high concentration of exporters, but a low concentration of importers. Few supply but many receive. Because so many countries are importers, many taking roughly similar amounts, it has not proved useful to try to put these countries into rank order, and they are listed alphabetically.

The countries have been arranged in order of the size their exports. This reveals Thailand to be the leading contemporary rice exporter, with over 5 million metric tons per year, nearly 30 per cent of the total. Then comes the US with approximately 2.5 million metric tons annually, roughly 15 per cent, and Vietnam much the same. Pakistan exports about 1.5 million metric tons each year, less than 10 per cent. India has normally exported about 0.6 million metric tons, about 4 per cent of world trade, but the volatility of the trade is revealed by their exceptional exports in 1995 of over 4 million metric tons, some 20 per cent of world trade, a situation examined in Chapter 5. Even in 1996 India exported 3.5 million tons, 18 per cent of the total. These Indian figures are highlighted (bold) in Tables 7.1 and 7.2. China too has demonstrated an erratic pattern, with exports of over a million tons in 1993 and 1994 and a mere 32,000 tons in 1995. As can be seen in the figures for imports in Table 7.3, China was a major

Table 7.1 World rice exports 1990–6, metric tons (000) (milled basis)

	1990	1991	1992	1993	1994	1995	1996
Thailand	3,938	3,988	4,776	4,798	4,738	5,931	5,280
USA	2,420	2,197	2,107	2,644	2,794	3,073	2,624
Vietnam	1,670	1,048	1,914	1,765	2,222	2,308	3,100
Pakistan	904	1,297	1,358	937	1,399	1,592	1,663
India	505	711	563	625	600	4,201	3,500
China	326	689	933	1,374	1,519	32	300
Australia	470	450	500	540	570	519	475
Uruguay	288	260	327	451	396	470	596
Burma	186	176	185	223	619	645	265
Others	954	1,243	1,419	1,558	1,608	2,228	1,522
Total	11,661	12,059	14,082	14,915	16,465	20,999	19,325

Source: Grain: World Markets and Trade.

Table 7.2 World rice exports 1990–6 (%)

	1990	1991	1992	1993	1994	1995	1996
Thailand	33.7	33.0	33.9	32.1	28.7	28.2	27.3
USA	20.7	18.2	14.9	17.7	16.9	14.6	13.5
Vietnam	14.3	8.6	13.5	11.8	13.4	10.9	16.0
Pakistan	7.7	10.7	9.6	6.2	8.4	7.5	8.6
India	4.3	5.8	3.9	4.1	3.6	20.0	18.1
China	2.7	5.7	6.6	9.2	9.2	0.1	1.5
Australia	4.0	3.7	3.5	3.6	3.4	2.4	2.4
Uruguay	2.4	2.1	2.3	3.0	2.4	2.2	3.0
Burma	1.5	1.4	1.3	1.4	3.7	3.0	1.3
Others	8.1	10.3	10.0	10.4	9.7	10.6	7.8

Source: *Grain: World Markets and Trade*.

importer that year, and we know that this is where much of India's rice ended up. Burma, once a major figure in the rice trade now exports less than 3 per cent of the total, and is usually surpassed by Australia. Even Uruguay normally exports more than Burma, although of course this rice is destined for other South American markets rather than Asia. So the big three these days are Thailand, US A. and Vietnam, together providing 50–60 per cent of the total. Pakistan, India and China are in secondary positions with in total 15–20 per cent normally, but India and China are prone to erratic fluctuations year by year as in 1995. But the high concentration of exports in the hands of these nine suppliers is indicated by the fact that some 90 per cent of world trade originates from them, leaving about 10 per cent only coming from other countries.

As can be seen very few countries take more than 5 per cent of world imports. Many only take 1–2 per cent. Of the larger importers Iran stands out usually taking in the range of 7 per cent, and Brazil is also a leading importer with 4–6 per cent. Brazil is mainly consumer of US and Uruguay rice. But the volatility of the rice trade is demonstrated by the figures for China, Japan and Indonesia. Perhaps Japan is the most startling case. After long pursuing a policy of rice self-sufficiency, 1993 saw a serious harvest shortfall, and in 1994 Japan suddenly imported 2.4 million metric tons (bold highlighted) some 15 per cent of the world's rice supply. The rice came from China and Thailand, even though these were not the qualities Japan consumers normally took. Fortunately the 1994 Japanese domestic harvest turned out to be excellent, and the imported rice was allowed to remain in storage. Later it was made available in food aid in which North Korea showed interest (see Chapter 6 page 78). Indonesia too has tried to pursue self-

Table 7.3 World rice imports 1990–6, metric tons (000) (milled basis)

	1990	1991	1992	1993	1994	1995	1996
Bangladesh	0	24	15	0	175	1,566	700
Brazil	493	772	456	831	1,098	987	800
China	57	67	93	112	700	1,964	850
Cote d'Ivoire	263	169	309	384	187	387	300
Cuba	238	264	198	397	252	316	400
EU	500	481	480	444	725	820	800
Indonesia	77	192	650	22	1,120	3,000	1,250
Iran	850	565	1,195	1,161	645	1,633	1,400
Iraq	388	252	448	647	64	92	250
Japan	11	34	17	107	2,473	29	450
Korea, North	27	194	10	112	53	683	350
Malaysia	298	367	468	385	317	402	550
Mexico	148	173	377	275	242	245	315
Nigeria	224	296	440	382	300	450	500
Peru	233	340	359	336	220	258	400
Philippines	538	91	6	215	0	277	900
Russia	100	100	500	127	48	125	350
Saudi Arabia	547	533	760	859	698	615	750
Senegal	332	433	333	396	252	402	700
South Africa	295	360	360	431	402	634	600
Sri Lanka	139	208	330	267	39	25	300
Turkey	203	146	313	309	235	445	350
UAE	317	249	65	75	80	85	85
USA	151	163	172	199	244	221	268
Yemen	150	111	169	131	172	68	100
Others	5,082	5,475	5,559	6,311	5,724	5,270	5,607
Total	11,661	12,059	14,082	14,915	16,465	20,999	19,325

Source: Grain: World Markets and Trade

Table 7.4 World rice imports 1990–6 (%)

	1990	1991	1992	1993	1994	1995	1996
Bangladesh	0.0	0.1	0.1	0.0	1.0	7.4	3.6
Brazil	4.2	6.4	3.2	5.5	6.6	4.7	4.1
China	0.4	0.5	0.6	0.7	4.2	9.3	4.3
Cote d'Ivoire	2.2	1.4	2.1	2.5	1.1	1.8	1.5
Cuba	2.0	2.1	1.4	2.6	1.5	1.5	2.0
EU	4.2	3.9	3.4	2.9	4.4	3.9	4.1
Indonesia	0.6	1.5	4.6	0.1	6.8	14.2	6.4
Iran	7.2	4.6	8.4	7.7	3.9	7.7	7.2
Iraq	3.3	2.0	3.1	4.3	0.3	0.4	1.2
Japan	0.0	0.2	0.1	0.7	15.0	0.1	2.3
Korea, North	0.2	1.6	0.0	0.7	0.3	3.2	1.8
Malaysia	2.5	3.0	3.3	2.5	1.9	1.9	2.8
Mexico	1.2	1.4	2.6	1.8	1.4	1.1	1.6
Nigeria	1.9	2.4	3.1	2.5	1.8	2.1	2.5
Peru	1.9	2.8	2.5	2.2	1.3	1.2	2.0
Philippines	4.6	0.7	0.0	1.4	0.0	1.3	4.6
Russia	0.8	0.8	3.5	0.8	0.2	0.5	1.8
Saudi Arabia	4.6	4.4	5.3	5.7	4.2	2.9	3.8
Senegal	2.8	3.5	2.3	2.6	1.5	1.9	3.6
South Africa	2.5	2.9	2.5	2.8	2.4	3.0	3.1
Sri Lanka	1.1	1.7	2.3	1.7	0.2	0.1	1.5
Turkey	1.7	1.2	2.2	2.0	1.4	2.1	1.8
UAE	2.7	2.0	0.4	0.5	0.4	0.4	0.4
USA	1.2	1.3	1.2	1.3	1.4	1.0	1.3
Yemen	1.2	0.9	1.2	0.8	1.0	0.3	0.5
Others	43.5	45.4	39.4	42.3	34.7	25.0	29.0

Source: *Grain: World Markets and Trade*

sufficiency, and after low imports in the early 1990s, found in 1994 that they had to import 6 per cent of the world supply, with 14 per cent in 1995, and 6 per cent again in 1996 (highlighted). But China perhaps above all others shows intriguing features, for it is the only country to figure prominently in both the export and import lists. The figures for China are abstracted in Table 7.5.

This brings out the situation. In 1994 China exported 1.5 million metric tons, or 9 per cent of world exports, but in 1995 imported 1.9 million metric tons, 9 per cent of world imports! Clearly swings of plus or minus 9 per cent are substantial influences in the operation of world trade. This switchback effect is somewhat modified by the net figures, for China always imports some high quality rice whilst exporting low quality rice. On this basis China exported 8 per cent of

Table 7.5 China's net rice exports 1990–6, metric tons (000) (milled basis)

	1990	1991	1992	1993	1994	1995	1996
Exports	326	689	933	1,374	1,519	32	300
Imports	57	67	93	112	700	1,964	850
Exp–Imp	+269	+622	+840	+1,262	+819	–1,932	–550
World %	+2.3	+5.1	+5.9	+8.4	+4.9	–9.2	–2.8

Source: Table 7.1 and Table 7.3.

world exports in 1993, and imported 9 per cent in 1995, but the ebb and flow effect is still apparent. So in the 1990s China has been a source of considerable instability. When the sudden demands of Japan and Indonesia are added to the picture, the volatility of the world rice market is all too evident. Fortunately the emergence of India as a major exporter averted a major disaster. Another Asian country demonstrating instability is Bangladesh, usually a minor importer, but in 1995 a major importer, taking 1.5 million metric tons, some 7 per cent of world imports.

Looking at the general pattern of world imports, special comment needs to be made on the Middle East, especially Iran, Iraq and Saudi Arabia. Iran regularly consumed about 7 per cent of world trade, and on this basis was the world's leading consumer. Saudi Arabia too was an important consumer, with about 4 per cent of world consumption. Iraq also was an important consumer taking 3–4 per cent until political circumstances intervened. Together these countries normally take 10–15 per cent of world trade. West Africa is also another notable consumer, with Senegal, Cote D'Ivoire and Nigeria taking together 6–7 per cent of world trade. Elsewhere the situation in South America needs comment, for as previously noted Brazil regularly takes 4–6 per cent of world trade, sourcing most of their imports from the United States and Uruguay. Of less importance are Mexico, and Peru each of which take 1–2 per cent most of which again is obtained from the United States and Uruguay, although some Vietnamese and Thai rice is involved.

What the future years indicate for the rice trade is a question of some conjecture. Forecasts for the world rice trade in 1997 are as low as 17 million tons, a considerable reduction from the 1996 levels of 19 million tons, the second highest ever total (*Grain* 1997 May: 3).

APPENDIX

Addresses and information sources

London Rice Brokers' Association

The monthly *Circular* of market intelligence can be obtained from the Secretary, M. French, address as for Jackson Son & Co below. Members of the Brokers' Association are:

Jackson Son & Co (London) Ltd
4 St Georges Yard
Farnham
Surrey GU9 7LW
Tel: 01252 741400
Telex: 8952627
Fax: 01252 727677

Charles Wimble Sons & Co Ltd
Thrale House
44 Southwark Street
London SE1 1UN
Tel: 0171 378 8822
Telex: 886881
Fax: 0171 407 6988

Schepens & Co SA
Bredabaan 44-B7
2930 Brasschaat
Belgium
Tel: 32 3 645 6360
Telex: 31235
Fax: 32 3 645 7148

APPENDIX

Marius Brun Et Fils
36 Boulevard Emile Zola
Boite Postale 111
13631 Arles, France
Tel: 33 9096 3646
Telex: 420310
Fax: 33 9093 5074

1997 Chairman – Robert Brun – Marius Brun et fils.

The rice industry in Australia centres on Leeton, New South Wales, and enquiries should be directed to:

Ricegrowers' Co-operative Ltd
Head Office
Yanco Avenue
Leeton
New South Wales 2705
Tel: 069 530 411
Fax: 069 534 733

They publish a magazine called *Milling Around*, and an annual report.

The US Rice Federation was composed of the USA Rice Council, the Rice Millers' Association and the US Rice Producers' Group. Programmes and administrative duties for all three organisations are now conducted by USA Rice Federation staff. All enquiries should be sent to either of the main offices:

USA Rice Federation
4301 North Fairfax Drive, Suite 305
Arlington, VA 22203–1616
Tel: 703 351–8162
Fax: 703 351–8162

USA Rice Federation
6699 Rookin
Houston
TX 77074
PO Box 740123
TX 77274
Tel: 713 270–6699
Fax: 713 270–9021

APPENDIX

(*Rice Journal: International Rice Industry Guide* 1997 May: 6)

Key US companies are:

American Rice Inc.
16825 Northchase Drive
Suite 1600
Houston
Texas 77060
Tel: 713 873–8800
Fax: 713 872–1929

Comet Rice is the marketing division of American Rice, and can be found at:

Comet Rice Division
PO Box 398
Maxwell
CA 95955
Tel: 916 438–2265
Fax: 916 438–2990

Comet Rice Division
PO Box 891
Stuttgart
AR 72160
Tel: 501 673–1616
Fax: 501 673–1270

Other major names are:

Cargill Rice Milling
Highway 82 West
Cargill Road
PO Box 690
Greenville
MS 38702–0690
Tel: 601 334–6266
Fax: 601 378–0355

APPENDIX

Continental Grain Company
Rice Division
277 Park Avenue
New York
NY 10172
Tel: 212 207–5100
Fax: 212 207–2904

Louis Dreyfus Corporation
PO Box 35
New Madrid
MO 63869
Tel: 573 643–2241
Fax: 573 643–2682

Global Rice Corporation
1300 SW 5th Avenue
Suite 3018
Portland
Oregan 97201–5637

For further information see *Rice Journal* (1997) May (International Industry Guide). The address of *Rice Journal* is:

Rice Journal
3000 Highwoods Blvd
Ste.300
Raleigh NC 27604
Tel: 919–872–5040
Fax: 919–876–6531
Email: SpecC@ad.com

Also see:

Creed Market Report
800 Wilcrest, Suite 200
Houston
TX 77042
Tel: 713–782–3260
Fax: 713–782–4671
Email: www.creedrice.com

Michael Creed at the above address is also a broker.

APPENDIX

Also from the text:

Mid America Commodities Exchange (also the address for the Chicago Board of Trade)
141 West Jackson Boulevard
Chicago
Illinois 60604
Tel: 001 312 341 7955
Fax: 001 312 341 3027

United States Department of Agriculture (USDA)
Foreign Agricultural Service:
Grain: World Markets and Trade
US Dept of Commerce, Technology Administration
National Technical Information Service
5285 Port Royal Road
Springfield, VA 22161
USA

Thai Rice Exporters' Association
37 Soi Ngamduplee
Rama 4 Rd
Yannawa
Bangkok
Thailand 10120
Tel: 00 66 2872674–7
    00 66 2866630
Fax: 00 66 2872678

REFERENCES

Adas, M. (1974) *The Burma Delta: Economic Development and Social Change on an Asian Rice Frontier, 1852–1941*, Madison, WI: University of Wisconsin Press.

The Australian (1984) 5 January: 1, 4.

Barker, R., Herdt, R.W. and Rose, B. (1985) *The Rice Economy of Asia*, Washington DC: Rescources For The Future.

The leading modern study of rice in Asia, and essential reading. Not much on trade.

Black, C. (1996) *Historical Development of the NSW Rice Industry: The Introduction of Rice to the Murrumbidgee Irrigation Area*, Leeton: New South Wales Ricegrowers' Co-operative Ltd.

Booth, A. (1977) 'Irrigation in Indonesia. Part I', *Bulletin of Indonesian Economic Studies* 13: 33–74

Brandt, L. (1989) *Commercialization and Agricultural Development: Central and Eastern China, 1870–1937*, Cambridge: Cambridge University Press.

Bray, F. (1986) *The Rice Economies: Technology and Development in Asian Societies*, Oxford: Blackwell.

An interesting individual study, particularly good on historical aspects, but does not deal with trade.

Broehl, W.G. Jr (1992) *Cargill: Trading the World's Grain*, Hanover, NH: University Press of New England.

A massive work about this great trading company. Mainly about wheat, with only a few references to rice.

Chang, T.-T. and Li, C.-C. (1991) 'Genetics and Breeding', in Luh (1991) vol I, 23–101.

Cheng, S.-H. (1968) *The Rice Industry of Burma*, Kuala Lumpur: University of Malaya Press.

Creed Rice Market Report.

For address see Appendix.

Daily Telegraph (1997) 9 April: 16; 6 May: 14.

Dalrymple, D.C. (1986) *Development and Spread of High Yielding Rice Varieties in Developing Countries*, Washington DC: USAID.

REFERENCES

David, C.C. and Otsuka, K. (1994) *Modern Rice Technology and Income Distribution in Asia*, Boulder, CO: Lynne Rienner.

Dethloff, H.C. (1988) *A History of the American Rice Industry*, College Station, TX: Texas A&M Press.
The definitive history of the US rice industry.

Drewry Shipping Consultants (1988) *Grain: Seaborne Trade and Transport*, London: Drewry.

—— (1995) *Multi-Purpose Cargo Ships and Their Future Markets*, London: Drewry.

—— (1996) *Container Shipping: The South East Asian Market*, London: Drewry.

Fearnley, A.S. (1995) *World Bulk Trades*, Oslo.

Financial Times (1995) 12 January: 39.

—— (1997) 8 January: 24; 2 April: 31.

Food and Agriculture Organisation of the United Nations (FAO) (1955) *The Stabilization of the International Trade in Rice: A Report of Possible Measures*, Commodity Policy Studies No. 7 (August), Rome: FAO.

—— (1991) *Demand Prospects For Rice and Other Foodgrains in Selected Asian Countries*, Rome: FAO.
A useful examination of current rice consumption in some Asian countries, and discussion of possible future trends.

Furnivall, J.S. (1944) *Netherlands India: A Study of Plural Economy*, Cambridge: Cambridge University Press.
The classic general economic study of the Dutch East Indies. See also next item.

—— (1948) *Colonial Policy and Practice: A Comparative Study of Burma and Netherlands India*, Cambridge: Cambridge University Press.

Grain: World Markets and Trade. See USDA.

Grist, D.H. (1975, 1986) *Rice*, London: Longman, 5th and 6th edn.
A masterpiece from the British colonial agricultural service. Each edition adds new insights. Essential!

Hayami, A. and Tsubouchi, Y. (eds) (1989) *Economic and Demographic Development in Rice Producing Societies: Some Aspects of East Asian Economic History, 1500–1900*, Tokyo: Keio University.
A wide ranging collection of conference papers, prepared for conferees. Worth examining in full, as only a few of the papers appear in edited form in the next item.

—— (1990) *Economic and Demographic Development in Rice Producing Societies: Some Aspects of East Asian Economic History (1500–1900)*, Session B-3, Economic and Demographic Development in Rice Producing Societies, Proceedings Tenth International Economic History Congress: Leuven University Press. (*Studies in Social and Economic History*, ed. Herman Van der Wee, vol. 6.)

Headrick, D.R. (1988) *The Tentacles of Progress. Technology Transfer in the Age of Imperialism*, New York: Oxford University Press.

REFERENCES

Contains some useful details of dams and irrigation works in British India and Egypt.

Hill, R.D. (1977) *Rice in Malaya: A Study in Historical Geography*, Kuala Lumpur: Oxford University Press.

The key work on the history of rice in Malaya.

Huff, W.G. (1994) *The Economic Growth of Singapore: Trade and Development in the Twentieth Century*, Cambridge: Cambridge University Press.

Huke, R.E. (1982) *Rice Area by Type of Culture: South, Southeast, and East Asia*, Manila, Philippines: International Rice Research Institute.

Huke, R.E. and Huke, E.H. (1990) *Rice: Then and Now*, Manila, Philippines: International Rice Research Institute.

Hungerford, D. (1950) 'The Australian Rice Harvest', *World Crops*, December, 498–501.

Ingram, J.C. (1971) *Economic Change in Thailand 1850–1970*, Stanford, CA: Stanford University Press.

International Rice Research Institute (IRRI) (1994) *IRRI 1993–1994: Filling the World's Rice Bowl*, Los Banos, Laguna, Philippines: International Rice Research Institute.

Essentially an annual report, giving recent developments.

Kito, H. (1989) 'History and Structure of Staple Foods in Japan', in Hayami and Tsubouchi (1989) 41–71.

Kose, H. (1994) 'Chinese Merchants and the Chinese Inter-Port Trade', in Latham and Kawakatsu (1994b) 129–44.

Latham, A.J.H. (1981) *The Depression and the Developing World, 1914–39*, London: Croom Helm.

—— (1986) 'The International Trade in Rice and Wheat Since 1968: A Study in Market Integration', in Fischer, W., McInnis, R.M. and Schneider, J. (eds) *The Emergence of a World Economy*, Part II, *1850–1914*, Wiesbaden GmbH: Franz Steiner Verlag, 645–63.

—— (1988) 'From Competition to Constraint: The International Rice Trade in the Nineteenth and Twentieth Centuries', *Business and Economic History* 17, 91–102.

—— (1994) '"Rice Moves to Areas Where Incomes Are Rising". A Re-Intepretation of Asian Economic Development since 1900', in Latham and Kawakatsu (1994a) 17–26.

Latham, A.J.H. and Kawakatsu, H. (eds) (1994a) *The Evolving Structure of the East Asian Economic System since 1700: A Comparative Analysis*, Milan: Università di Bocconi. (Proceedings Eleventh International Economic History Congress, Session B-6).

Milan conference papers from 1994 on Asian economic development, some of which deal with rice. Not to be confused with next item.

—— (1994b), *Japanese Industrialisation and the Asian Economy*, London: Routledge.

Leuven conference papers from 1990 on Asian economic development. Some papers deal with rice.

REFERENCES

Latham, A.J.H. and Neal, L. (1983) 'The International Market in Rice and Wheat, 1868–1914', *Economic History Review* 34, 260–80.

Lee, H. (1989, 1990) 'Rice Culture and Demographic Development in Korea, c.1429–1918', in Hayami and Tsubouchi (eds) (1989) 21–30, (1990) 55–71.

Lee, T.-W. (1996) *Shipping Developments in the Far East: The Korean Experience*, Aldershot: Avebury.

Lewis, G. (1994) *An Illustrated History of the Riverina Rice Industry*, Leeton: New South Wales Ricegrowers' Co-operative Ltd.

London Rice Brokers' Association Circular.
> A monthly circular giving market intelligence. For address see Appendix.

Luh, B.S. (1991) *Rice*, 2nd edn, vol. I, *Production*, vol. II, *Utilization*, New York: Von Nostrand Reinhold.
> A collection of papers on most aspects of the United States rice industry other than trade.

Luh, B.S. and Mickus, R.R. (1991) 'Parboiled Rice', in Luh (1991) vol. II, 51–88.

McCaughey, P. (1955) *Samuel McCaughey – A Biography by Patricia McCaughey*, Australia: Ure Smith Pty. Ltd.

Mikkelsen, D.S. and De Datta, S.K. (1991) 'Rice Culture', in Luh (1991) vol. I, 103–86.

Morgan, D. (1979) *Merchants of Grain: The Power and Profits of the Five Great Companies at the Centre of the World's Grain Supply*, New York: Viking Press.
> A journalistic account, with some sensationalised references to rice. Closer to fiction than fact!

Owen, N.G. (1984) *Prosperity Without Progress: Manila Hemp and Material Life in the Colonial Philippines*, Berkeley, CA: University of California Press.

Reynolds, L.G. (1985) *Economic Growth in the Third World, 1850–1980*, New Haven, CT: Yale University Press.
> A substantial compendium with much on agricultural policy and development.

Rice Journal
> The journal of the US rice industry. See 1997, May (International Industry Guide); 1997, 15 March, 24–8,'Vietnam Catching Up With The World'. See Appendix for address.

Robequain, C. (1944) *The Economic Development of French Indo-China*, London: Oxford University Press.
> Still the best study of French Indo-China, with much on rice and irrigation.

Roche, J. (1992) *The International Rice Trade*, Cambridge: Woodhead Publishing.
> An interesting study well worth consulting.

REFERENCES

Sauer, J.D. (1993) *Historical Geography of Crop Plants: A Select Roster*, Boca Raton, FL: C.R.C. Press.
Contains a useful short survey of rices and their distribution 206–12.

Sewell, T. (1992) *The World Grain Trade*, Hemel Hempstead: Woodhead-Faulkner.
Mainly about wheat, but some comments on rice.

Siamwalla, A. and Haykin, S. (1983) *The World Rice Market: Structure, Conduct and Performance*, Research Report 39 (June), Washington DC: International Food Policy Research Institute.
This report is crucial to our understanding of the world rice market.

Swaminathan, M.S. (1984) Rice, *Scientific American* 250, 62–71.
Essential reading on the evolution of rice in past and recent times.

Tsujii, H. (1977a) 'An Economic and Institutional Analysis of the Rice Export Policy of Thailand: With Special Reference to the Rice Premium Policy', *The Developing Economies* XV-2, 202–20.

—— (1977b) 'Rice Economy and Rice Policy in South Vietnam up to 1974', *South East Asian Studies* 15, 263–94.

—— (1995) 'Characteristics of and the Trade Conflicts in the International Rice Market: A Case Against the Free Trade Postulate', *The Natural Resource Economic Review* 1, 119–35.

United States Department of Agriculture (USDA) *Grain: World Markets and Trade*.
This is published monthly by the USDA Foreign Agricultural Service, and contains useful statistical information and country surveys. For address see Appendix.

van der Eng, P. (1994) 'Development of Seed-Fertilizer Technology in Indonesian Rice Agriculture', *Agricultural History* 68, 20–35.

—— (1996) *Agricultural Growth in Indonesia: Productivity Change and Policy Since 1880*, Basingstoke and London: Macmillan.

Vergara, B.S. (1991) 'Rice Plant Growth and Development', in Luh (1991) vol. I, 13–22.

Wadsworth, J.I. (1991) 'Milling', in Luh (1991) vol. I, 347–88.

Webb, B.D. (1991) 'Rice Quality and Grades', in Luh (1991) vol. II, 89–119.

Wickizer, V.D. and Bennett, M.K. (1941) *The Rice Economy of Monsoon Asia*, Stanford, CA: Food Research Institute.
This study of the rice economy of Asia between the two great wars is still important reading.

INDEX

Note: page references for major entries are shown in **bold**

Africa 3–5, 22, 25, 98; trade in rice 42, 49, 60–2, 73–4, 84, 87; *see also* East Africa; North Africa; West Africa
agricultural research stations 4
airborne sowing 17–18
American Rice, Inc. 38, 65, 84, 101
Amran, Salvadore 39
Andre (company) 38
Anhui 74–5
Arbitration Acts (UK) 39
Arequippa 92
Argentina **91**; trade in rice 88, 89–90, 92
Arkansas 17, 41, 85
Arrocera Argentina, La (company) 91
Asia *see* East Asia; South Asia; South-East Asia
Asian Development Bank 65
Australia 6, 17, **82–3**; trade in rice and government policies 77, 78, 81, 82–3, 94–5
azolla (water fern) 8

Balfour Maclaine (company) 38
Bangkok 28, 61; brokers and traders 36–7, 41, 42
Bangladesh **51**; irrigation and cultivation 10, 15; trade in rice and government policies 33–4, 49, 50, 51, 53, 62, 96–8; *see also* Pakistan
Bank of Thailand 61
banks 47, 52, 54, 61, 65
Barker, R. *et al.*: on brokers and traders 35; on government policies 48, 55, 59, 61, 68, 73, 79, 81; on irrigation and cultivation 11, 14, 15–16; on milling 21; on trade in rice 28, 30, 32, 34; on varieties of rice 6, 8–9
basmati rice 21, 50, 89
Batavia 28
bean cake 14
Belgium 27, 99
Bengal 28, 46
Bennett, M.F. 30, 35, 55, 59, 81
Bhumipo reservoir 60
BIMAS (Indonesia) 56, 57, 58
Black, C. x, 83
Blue Belle variety 90, 92
Bombay 43
Booth, A. 15
boro rice 51
bran removal *see* milling
Bray, F.: on irrigation and cultivation 10, 11, 12, 13–14; on trade in rice 28; on varieties of rice 4, 5, 8
Brazil **89–90**; and government policies 88–90, 92; trade in rice 39, 95, 96–7, 98
breeding experiments 6–7

INDEX

Britain and British colonialism: brokers and traders 36, 39, 40–1, 99; and government policies 46, 50, 51, 54; irrigation and cultivation 13, 14; trade in rice 27–8, 40–1, 84

Broehl, W.G., Jr 39

brokers and traders 27–8, **35–45**, 99–100; *see also* trade and commerce

Brun, Marius, Et Fils (company) 100

Bulloch Bros. (company) 39

BULOG (National Logistics Agency, Indonesia) 57–9

Burma *see* Myanmar

California 17, 29, 82, 85, 86, 87

Cambodia 10, 13, 15, 30

Canada 77, 87

CandFFo (cost and freight free out) basis 36

Capital Rice Group (company) 37

Cargill (ex Tradax, company) 36, 37–8, 78, 101

cargo rice (*loonzain*) 23

Caribbean and Central America 25, 28, 38, 42; trade in rice and government policies 65, 73–4, 84, 88

Castro, F. 84

CCC (Commodity Credit Corporation, USA) 83, 85

Central America *see* Caribbean and Central America

Cereals, Oils and Foodstuffs Import and Export Corporation (China) 74–5; *see* CEROIL

CEROIL (company) 74–6

Chaiyaporn (company) 37

'Champa' varieties 4

Chang, T.-T. 5, 6, 7, 9, 15

Chao Phraya River and delta 28, 36, 60

chemicals *see* fertilisers; herbicides and pesticides

Chicago 28, 29, 103; Rice and Cotton Exchange 40–1

Chile 91

China **68–78**; and government policies 53, 57, 61–3, 76, 78, 81–2, 84, 87; irrigation and cultivation 14, 15, 17; milling 20–1; varieties of rice 2, 3, 4, 5, 6, 8, 9; *see also* China and trade

China and trade in rice 28–34; in 1990s 93–8; brokers and traders 29, 38, 42–3; East Asia, rest of 78, 81, 82; South-East Asia 53, 57, 61–2, 63; United States 76, 84, 87

Chinese workers abroad 28, 36–7, 85

China National Liang Feng Grain Import and Export Corporation 76

Cholon 67

CIC (company) 40

collectivisation 62, 69

colonialism 13–15; and government policies 46, 50, 51, 52, 54, 55; *see also* Britain; Netherlands

Comet Rice (company) 38, 101

Commodity Credit Corporation (USA) 83, 85; *see also* CCC

Community Development Programme and Organisation (India) 46

Continental Grain Company 36, 37–8, 39, 102

contract system in Vietnam 62–3

cost and freight free out basis 36; *see also* CandFFo

Cote D'Ivoire *see* Ivory Coast

Creed Rice (company) 36, 38–9, 45, 102

Cuba 42, 65, 73–4, 84

cultivation *see* irrigation and cultivation

Cultivation Extension *see* OPSUS

Cultural Revolution 69

Czechoslovakia 84

Dakar 4, 53

dams and reservoirs *see under* rivers

Dar es Salaam 42

Darwin 83

110

INDEX

David, C.C. 6, 15
De Datta, S.K. 10
Dee-geo-woo-gen variety 6
dehusking 22, 23–5
deltas *see* rivers and deltas
DEMAS (Indonesia) 56
Department of Agriculture (USA) 81, 84–5, 103
depression, 1930s economic 30, 54, 55, 67, 80
Dethloff, H.C. 4, 84
Development Plans: Bangladesh 51; China 72–3
diseases of rice 8
Doi Moi policy reforms 64
Dojima rice Exchange 40
Dreyfus, Louis, Corporation 38, 102
dry cultivation 3, 9, 10, 82, 89–90
drying 23, 25
Dutch East Indies *see* Indonesia
Dutch Guiana (Surinam) 17
dwarf and semi-dwarf varieties *see* hybrid varieties

East Africa 4, 25, 42
East Asia *see* China; Japan; Korea; Taiwan
Eastern Europe 73–4, 75, 84
EC 38, 84
ECASA (Peru) 91
eleusine coracana (ragi) 3, 34
Entre Rios 91
Europe 4; brokers and traders 38, 39–40, 41, 99, 100; Eastern 73–4, 75, 84; and government policies 77, 84, 87; and irrigation and cultivation 14, 15; trade in rice 27–8, 77
Export Guarantee Programme (USA) 85
Export and Import Services, Myanmar 53
exports *see* trade and commerce

F1 hybrid rices 9, 69–70
FA (Food Agency, Japan) 80–1, 82
Fair Price shops (India) 48

famine and starvation 46, 47, 51, 69, 78; alleviation *see* subsidies
FAO *see* Food and Agriculture Organisation of the United Nations
farming *see* irrigation and cultivation
FAS (Foreign Agricultural Service, USA) 84, 85
FAS (free along side) 41, 45
Feather River 17
fertilisers: and irrigation 12, 13, 14, 15, 16, 31; manure and compost 8, 14; and varieties of rice 6–8; *see also* fertilisers and government policies
fertilisers and government policies: East Asia 68, 69, 70, 72, 73; South-East Asia 47, 50, 52, 53, 55, 56, 57, 58, 60, 62, 67
fish 9
'floating' deep-water cultivation 3, 10
Food Agency (Japan) 80–1, 82
Food, Agriculture, Conservation and Trade Act (USA) 85
Food and Agriculture Organisation of the United Nations 31, 34; and government policies 48, 51, 59, 68, 73, 75
Food Corporation (India) 42, 47, 48
food distribution *see* subsidies
'Food for Peace' (USA) 83–4
Foodgrain and Edible Oil Bureau, State (China) 69, 70, 71
Ford Foundation 5
Foreign Agricultural Service (USA) 84, 85
France 14, 15, 27, 39, 100; *see also* Indochina
free along side (FAS) 41, 45
free on board (FOB) 36, 41, 43, 44, 87
free food *see* subsidies
Fujian 9
fungus 8
Furnivall, J.S. 15
Futures and Options Exchange, London 40–1

INDEX

Ganges River 14
GATT (General Agreement on Tariffs and Trade) 78, 79, 81, 82, 87
Glencore Denrees (ex Richco, company) 36, 37–8, 40
Global Rice Corporation 40, 102
glutinous rice 20–1, 40
Golden Resource Corporation 65
Goletti, Francesco 66
government policies and countries 29–30, 35–8, **46–92**; *see also* Australia; East Asia; Latin America; South Asia; South-East Asia; United States
grades and standards of rice 20–2, 44–5
grain size 20–1, 79, 82, 85–7
Green Revolution 6, 7, 9, 14, 31, 67
Grist, D.H.: on government policies 83; on irrigation and cultivation 10–12, 13, 17–19; on milling 22–3, 24–6; on trade in rice 34; on varieties of rice 3, 5, 7–8, 9
GSM-102 (Export Guarantee Programme, USA) 85
GSM-103 (Intermediate Export Credit Guarantee Programme, USA) 85
Guangdong 9
Gulf Pacific Rice Co. 38
Guyana 25

Hainan Island 9, 69
Haiti 88
Hanoi 63, 65
Harper, Alan x, 45
harrowing 12–13, 17
harvesting 6, 13, 18
Hawaii 29
Haykin, S. 20–1, 27–8, 30, 31–2
Headrick, D.R. 4, 14
herbicides and pesticides 8–9, 13, 15, 16, 56, 57
High Yielding Varieties *see* HYVs
Hill, R.D. 14
Ho Chi Minh City (Saigon) and trade in rice 28, 29, 67; brokers and traders 42, 44; and government policies 63, 65
Hong Kong and trade in rice 27, 28–9, 32; brokers and traders 35–6; East Asia 68, 73, 74; South-East Asia 61; United States 87
household contract system in China 70–2, 73
Houston 41, 85
Hubei 74–5
Huke, R.E. and E.H. 10, 15, 19
hulling *see* milling
Humpty Doo 83
Hunan 9, 43, 70, 74–5
Hungary 84
Hungerford, D. 83
Hwang He River 32
hybrid varieties 6, 7, 8, 9, 52, 57, 69–70, 73, 80; *see also* HYVs
HYVs (High Yielding Varieties) 31; East Asia 69, 70; South Asia 47, 48, 50; South-East Asia 52, 53, 57, 60, 62, 63, 67, 69–70

imports of rice 32–4, 50, 77, 90; controls 29; East Asia 74, 79, 80, 81–2, 85–6; Latin America 88–9, 90–2; Middle East 61, 84, 87; South-East Asia 55, 56, 57, 58, 59, 62, 67–8, 77–9, 80–1, 86–7; United States 89; *see also* subsidies; trade and commerce
India **46–50**; and government policies 46–50, 63, 77, 84, 86, 89; irrigation and cultivation 14, 16–17; milling 24, 25; varieties of rice 2, 5, 7; *see also* India and trade
India and trade in rice 32, 33; in 1990s 42–3, 93–5, 98; brokers and traders 42–3, 44; East Asia 77; South America 92; South-East Asia 63; United States 84, 86, 89
Indian workers abroad 25, 28
indica varieties 3–4, 6

INDEX

Indochina: irrigation and cultivation 10, 13, 14–15; trade in rice 28, 30–1, 33–4, 54, 55, 68; varieties of rice 4, 5, 8; *see also in particular* Cambodia; Laos; Thailand; Vietnam

Indonesia **55–9**; and government policies 49, 55–9, 60–2, 65, 73, 84, 86–7; irrigation and cultivation 10, 11, 14–15; varieties of rice 2, 3, 6; *see also* Indonesia and trade

Indonesia and trade in rice 28, 32, 33–4, 49, 73; in 1990s 95–7, 98; brokers and traders 35, 38, 42; South-East Asia, rest of 60, 61–2, 65; United States 84, 86, 87

INMAS (Indonesia) 56

insecticides *see* herbicides and pesticides

INSUS (Indonesia) 58

Intensive Agricultural District and Areas Programmes (IADP and IAA, India) 46–7

Intermediate Export Credit Guarantee Programme (USA) 85

International Rice Industry Guide 45, 102

International Rice Research Institute (Los Baños) 64; and government policies 67, 69; and irrigation and cultivation 10, 16, 17, 18; and varieties of rice 5–6, 7–8, 9; and yield of rice 29

International Rice Research Institute (Washington) 66

IR varieties 6–7, 9

Iran and trade in rice 49, 50, 61, 63, 64, 84, 91, 96, 97, 98

Iraq and trade in rice 36, 50, 65, 84, 96, 97, 98

Irrawaddy delta 28

IRRI *see* International Rice Research Institute (Los Baños)

irrigation and cultivation **10–19**; and government policies 47–8, 53–5, 60, 62, 67, 82–3, 90; and varieties of rice 3, 5, 6, 9; *see also* dams and reservoirs; rivers and deltas

Ivory Coast (Cote D'Ivoire) 74, 98

Jackson Son & Co. 36, 45, 99

Japan **80–2**; and government policies 78, 84, 85, 87; irrigation and cultivation 10, 13, 14, 15; milling 20; varieties of rice 2, 3, 4, 5, 6; *see also* Japan and trade

Japan and trade in rice 29, 32, 33–4, 78; in 1990s 95–7, 98; brokers and traders 35, 40; United States 84, 85, 87

japonica varieties 3–4, 79

Java 10, 14–15, 57

javonica varieties 3–4

Jiangsu 74

Jiangxi 43, 74–5

Jordan 87

Kakaku Kikoo (Japan) 81

Kakinada 43

Kandla 43, 49

keeping quality 26

Kerala 47

Kito, H. 3

Kohsichang 41

Korea **78–9**; and government policies 77–9, 81, 84–6; irrigation and cultivation 14, 15; milling 20; trade in rice 34, 74, 77–8, 81, 84–6, 95–7; varieties of rice 3, 4, 6; war 79

Kose, H. 14

Kyushu 2

Laguna de Merin area 90

Lake Charles 41, 85

Laos 13, 15, 21, 30, 81

Latham, A.J.H.: on brokers and traders 36, 39, 40; on government policies 46, 55, 59, 68; on trade in rice 28–9, 30, 34; on varieties of rice 2

Latin America and trade in rice 4, 63, 82, 87–8; in 1990s 94–5, 96–7, 98; brokers and traders

38, 39; and government policies 89–92; *see also* Argentina; Brazil; Caribbean; Peru; Uruguay
Le Van Triet 64
Lee, H. 3, 79
Leeton 83, 100
Lewis, G. 83
Li, C.-C. 5, 6, 7, 9, 15
Lima 92
London Futures and Options Exchange (FOX) 40–1
London Rice Brokers' Association and *Circular* 27–8, 36, 37, 39, 40, 41–5, 99; and government policies 50–1, 59, 60–1, 63, 77, 79, 81–2
loonzain (cargo rice) 23
Louisiana 17, 85
LRBA(C) *see* London Rice Brokers' Association and *Circular*
Luh, B.S. 26
Luzon 10–11

Madagascar 4, 25
Madras 14
MAFF (Ministry of Agriculture, Forestry and Fisheries, Japan) 80–1
Malaysia (and Malaya) 54–5; irrigation and cultivation 14, 15, 16; milling 25; varieties of rice 2; *see also* Malaysia (and Malaya) and trade
Malaysia (and Malaya) and trade in rice 28, 30–4; in 1990s 96–7; brokers and traders 35, 36; and government policies 54–5, 61, 64
manure and compost 8, 14
Mao Zedong 70
Marc Rich Investments (company) 39
Mauritius 25, 53
mechanisation 23; and irrigation and cultivation 13, 16–17, 18–19, 90
Mekong River and delta 10, 14, 15, 28, 62, 65
MEP (minimum export price) 63

MercoSur free trade area 90
Mexico 88–9, 96–7
Mickus, R.R. 26
Mid America Commodities Exchange 40, 103
Middle Eastern trade in rice (and government policies) 4, 21; brokers and traders 36, 38; South America 91; South Asia 49, 50; South-East Asia 60, 61, 63, 64, 65; United States 84, 87
Mikkelsen, D.S. 10
Millet 2, 3, 34
milling 20–6; and government policies 63, 65, 66, 81, 88, 89, 91, 92
minimum export price 63
Ministries: of Agriculture, Forestry and Fisheries (Japan) 80–1; of Agriculture and Irrigation (Myanmar) 53–4; of Agriculture and Rural Development (Vietnam) 64, 65; of Commerce (China) 74; of Commerce (Thailand) 61; of Community Development (India) 46; of Foreign Economic Relations and Trade (MOFERT, China) 74, 75; of Internal Trade (China) 75–6; of Planning and Investment (Vietnam) 64; of Trade (Vietnam) 64; *see also* Department of Agriculture (USA)
MIT (Ministry of Internal Trade, China) 75–6
MOFERT (China) 74, 75
Mogul Street (Rangoon) 40
Mombasa 42
monsoons 12, 42
Morgan, D. 37, 39, 41
Muda River irrigation System 15, 54
Murray River district 82
Murrumbidgee Irrigation Area 17, 82–3
Myanmar 52–4; irrigation and cultivation 10, 15; milling 21,

INDEX

23; varieties of rice 5; *see also* Myanmar and trade
Myanmar and trade in rice 28, 30, 31; in 1990s 94–5; brokers and traders 42; and government policies 46, 49, 52–4, 59, 74

NAFTA 88
National Logistics Agency (Indonesia) 57–9
Neal, L. 29, 40, 46
Nepal 49, 81
Netherlands 13, 14–15, 55
New Orleans 28, 40, 41, 85
New Orleans Commodity Exchange 28
New South Wales 17, 82–3
Nguyen Dang Chi 66
Nigeria 36
nitrogen fixation 8
North Africa 4, 84
North American Free Trade Area 88
North Korea 6, **78**; and government policies 74, 78, 81; irrigation and cultivation 14, 15; milling 20; and trade in rice 74, 81, 95–7; varieties of rice 3, 4, 6
Northern Food Corporation (earlier Vinafood I) 63, 64, 65, 66
Northern Territory (Australia) 83
nursery cultivation 13

Oakland 85
Office of Supply (South Korea) 79
oil price rise crisis 57, 60
oils added 24
OPSUS (Indonesia) 58
ORCO (company) 40
Oryza: O. glaberrima 3, 5; *O. nivara* 7; *O. sativa* 3
Osaka 40, 80–1
OSROK (South Korea) 79
Otsuka, K. 6, 15
output *see* yield

paddy, definition 22; milling quality of 22–3

Pakistan 21, **50–1**; and government policies 49, 50–1, 59, 77, 82, 84, 89; *see also* Pakistan and trade
Pakistan and trade in rice 28, 30, 31, 33, 49; in 1990s 93–5; East Asia 77, 82; South America 92; South-East Asia 59; United States 84, 89; *see also* Bangladesh
Pampanga River Project 15
parboiled rice 21, 22–3, 25–6, 43, 44
Paso-144 variety 90, 91
payment-in-kind 83
pearling 24
Pelita variety 57
Penang 28
Perkins, M. x, 64
Peru 38, 88, **91–2**, 98
pesticides *see* herbicides and pesticides
pests 7, 8, 9
Peta variety 6
Philippines **67–8**; and government policies 57, 59, 65, 67–8, 86, 87; irrigation and cultivation 10, 15, 16; varieties of rice 5, 6, 7; *see also* Philippines and trade
Philippines and trade in rice 29–34; in 1990s 96–7; brokers and traders 35, 38, 42; South-East Asia, rest of 57, 59, 65; United States 86, 87
PL480 programme (USA) 47, 83, 84, 85–6, 87
ploughing and harrowing 12–13, 17
Poland 84
policies *see* government policies
polishing 24–5
poverty alleviation *see* subsidies
prices 20–1, 29, 30, 43–4; East Asia 69, 71–5, 80–1; South Asia 47, 48–9, 50; South-East Asia 52, 53, 55–8, 60–1, 63
private brokerage companies 36, 38–9
production *see* yield

115

INDEX

Public Distribution System (PDS, India) 48
Public Food Distribution System (PFDS, Bangladesh) 51
Public Food Distribution System (PFDS, China) 71–2
Punjab 14, 21

ragi (*eleusine coracana*) 3, 34
rain-fed rice *see* dry cultivation
Rangoon (Yangon) 28, 29, 40, 42
Real Trading (company) 40
reservoirs *see* dams and reservoirs
responsibility system in China 70–2, 73
Reynolds, L.G. 50, 51, 53, 61, 73
rice *see* brokers and traders; government policies; irrigation and cultivation; milling; trade and commerce; varieties
Rice Council for Market Development (USA) 84, 85, 89, 100
Rice Export Corporation of Pakistan 50–1
Rice Federation, USA 89, 100
Rice Journal: address 102
Rice Millers' Association (USA) 81, 89, 100
Rice Price Advisory Council (Japan) 81
Rice Producers' Group (USA) 89, 100
Ricegrowers' Co-operative Ltd (Australia) 100
Richco (company) *see* Glencore Denrees
Rio Grande Do Sul 90
Riverina district 82, 83
rivers and deltas: Australia 82, 83; Brazil 90; China 2, 4, 28, 32; dams and reservoirs 12, 16, 60; India 14; Malaysia 28, 54; Mayanmar 28; Thailand 28, 36, 60; United States 17; Vietnam 10, 14, 15, 28, 62; *see also* irrigation and cultivation
Riz et Denrees (company) 39
Robequain, C. 14

Roche, J.: on trade in rice: brokers and traders 36, 39, 40, 41; East Asia 73, 75, 79, 81; South Asia 48, 50; South-East Asia 53, 55, 59, 61, 63; United States 85
Rockefeller Foundation 5
Romania 84
RPAC (Rice Price Advisory Council, Japan) 81

Sacramento River 17
saffron added 24
SAGR (China) 75–6
Saigon *see* Ho Chi Minh City
San Joaquin River and valley 17
Sao Paulo 91
Saudi Arabia 49, 84
Schepens & Co SA 99
sea, transport by *see* shipping; trade and commerce
self-sufficiency goal 30, 31, 32; and government policies 51, 54, 58, 67, 68, 70, 79, 81
Senegal 43, 49, 62, 91
Sewell, T. 36, 37, 39, 40, 41
SFC *see* Southern Food Corporation
shadoof 16–17
sheath blight 8
shipping 29, 41–3
Siamwalla, A. 20–1, 27–8, 30, 31–2
Sichuan 9
Singapore and trade in rice 27, 28, 29, 32; brokers and traders 35–6; and government policies 61, 64, 87
sinica-japonica varieties 3–4
Sirikit reservoir 60
slash-and-burn farming 10
smuggling 53, 63
Songwad Road (Bangkok) 41
Soon Hua Seng Co. 37
South Africa 84, 96–7
South America *see* Latin America
South Asia *see* Bangladesh; India; Pakistan; Sri Lanka
South Korea 79; irrigation and cultivation 14, 15; milling 20;

INDEX

trade in rice 34; varieties of rice 3, 4, 6
South-East Asia *see* Cambodia; Indonesia; Laos; Malaysia; Myanmar; Philippines; Thailand; Vietnam
Southern Food Corporation (earlier Vinafood II) 63, 64–5, 66
Soviet Union, former 63, 84
sowing seed 17–18
Sri Lanka 10, 25; and trade in rice 28, 31, 33, 53, 74, 96–7
standards *see* grades
starvation *see* famine
State Administration of Grain Reserve (SAGR, China) 75–6
Steels (company) 39
subsidies and poverty alleviation: China 69, 70–2; food distribution 48, 71–2; Myanmar 52; North Korea 78; South Asia 48, 50, 51; *see also* imports
Sucden (company) 40
Suharto, General 56
Sukarno, President 56, 57
Sumatra 57
SUPRA INSUS (Indonesia) 58
Surinam (Dutch Guiana) 17
Swaminathan, M.S. 4–6, 7, 8, 9

Taichung Native 1 variety 6, 7
Taiwan **79–80**; and government policies 79–80; irrigation and cultivation 13, 14, 15; milling 20; varieties of rice 6, 7
Takasuka, I. 82
Territory Rice Ltd (Australia) 83
Texas 17, 85
Thailand 15, **60–2**; and government policies 49, 53, 54, 55, 57, 59, 60–2, 68, 86, 89, 91, 92; milling 20, 21; *see also* Thailand and trade
Thailand and trade in rice 28–31, 49, 53, 54, 55, 57, 59, 86, 89, 91; in 1990s 92, 94–5; brokers and traders 36–8, 40–4; East Asia 74, 77, 78, 81, 82; and government policies 61–2, 68,
92; Thai Rice Exporters Association 36–7, 61–2, 103, United States 86, 87
Tokyo 80–1
Totah, Raphael 39
Tradax *see* Cargill
trade and commerce **27–34**; in 1990s **93–8**; and milling 20, 21; *see also* brokers and traders; government policies; imports; prices; and under individual countries
traders and brokers, difference between 36; *see also* brokers and traders
Training and Visits System (India) 47
Tsujii, H. 60, 62
Turkey 87, 91, 96, 97
turmeric added 24

United Arab Emirates 49
United Nations *see* FAO
United Nations and North Korea 78
United States **83–9**; and government policies 59, 60, 77, 78, 81–2, 83–9, 91, 92; irrigation and cultivation 17, 18; milling 20, 21–3, 26; varieties of rice 4, 5–6; *see also* United States and trade
United States and trade in rice 27–31; in 1990s 93–5, 96–7, 98; addresses 100–3; brokers and traders 36, 38–45; East Asia 76, 77, 78, 81–2, 84, 85, 86, 87, 88, 89; South America 89, 91, 92; South Asia 84, 86, 89; South-East Asia 59, 60, 84, 86, 87
upland cultivated rice *see* dry cultivation
Uruguay **90–1**; Round of GATT 78, 79, 81, 82, 87; trade in rice and government policies 82, 89–92, 94–5, 98
USDA *see* Department of Agriculture (USA)

INDEX

Van der Eng, P. 13
varieties of rice 2–9, 15, 16, 79, 90, 92; *see also* hybrid varieties
Vergara, B.S. 13
Vietnam 62–6; and government policies 62–6, 73, 77, 84, 86, 91, 92; irrigation and cultivation 15; milling 21; varieties of rice 5, 8; war 62, 73; *see also* Vietnam and trade
Vietnam and trade in rice 28–34; in 1990s 93–5; brokers and traders 38, 42, 44; East Asia 73, 77; South America 91, 92; United States 84, 86
village centres (Indonesia) 57
Vinafood I *see* Northern Food Corporation
Vinafood II *see* Southern Food Corporation

Wadsworth, J.I. 22, 23, 24, 26
Waller, Douglas x, 42
water *see* irrigation; rivers and deltas
water fern (*azolla*) 8
waterwheel 16
Webb, B.D. 20, 22, 23
weeding 13

West Africa: and government policies 60, 74; trade in rice 36, 38, 43, 60, 74, 98; varieties of rice 3–5
West Indies *see* Caribbean
wet cultivation *see* irrigation
wheat as alternative 29, 30, 32, 34, 51, 68, 81
'Whole Township Programme' (Myanmar) 52–3
Wickizer, V.D. 30, 35, 55, 59, 81
Wimble, Charles, Sons and Co.Ltd 39, 99
winnowing 22, 23
World Bank 47, 52, 54
World Food Programme 78
World Trade Organisation 76

Yangon *see* Rangoon
Yangzi River and delta 2, 4, 28, 32
Yellow River 32
yield 13–14, 15, 47; East Asia 70–1, 72; South-East Asia 52–3, 54–5, 56–7, 58, 67–8
Yugoslavia, former 84

Zhong Gu National Grain Enterprise Holding Corporation 76